T0073090

Creating Precision Robots

Creating Precision Robots
A Project-Based Approach to the Study of Mechatronics and Robotics

Francis Nickols
The University of Nottingham Ningbo China, Ningbo, China

Yueh-Jaw Lin
The University of Nottingham Ningbo China, Ningbo, China

Butterworth-Heinemann
An imprint of Elsevier

Butterworth-Heinemann is an imprint of Elsevier
The Boulevard, Langford Lane, Kidlington, Oxford OX5 1GB, United Kingdom
50 Hampshire Street, 5th Floor, Cambridge, MA 02139, United States

Notices
Knowledge and best practice in this field are constantly changing. As new research and experience broaden our understanding, changes in research methods, professional practices, or medical treatment may become necessary.

Practitioners and researchers must always rely on their own experience and knowledge in evaluating and using any information, methods, compounds, or experiments described herein. In using such information or methods they should be mindful of their own safety and the safety of others, including parties for whom they have a professional responsibility.

To the fullest extent of the law, neither the Publisher nor the authors, contributors, or editors, assume any liability for any injury and/or damage to persons or property as a matter of products liability, negligence or otherwise, or from any use or operation of any methods, products, instructions, or ideas contained in the material herein.

Library of Congress Cataloging-in-Publication Data
A catalog record for this book is available from the Library of Congress

British Library Cataloguing-in-Publication Data
A catalogue record for this book is available from the British Library

ISBN: 978-0-12-815758-9

For information on all Butterworth-Heinemann publications visit our website at https://www.elsevier.com/books-and-journals

 Working together
to grow libraries in
developing countries

www.elsevier.com • www.bookaid.org

Publisher: Katey Birtcher
Acquisition Editor: Steve Merken
Editorial Project Manager: Susan Ikeda
Production Project Manager: Vijayaraj Purushothaman
Cover Designer: Matthew Limbert

Typeset by SPi Global, India

Contents

Preface

AN OVERVIEW OF WHAT THIS BOOK DOES

This book teaches you the theory and practice of creating three precision robots which are:

The hitting robot with 10-ball magazine autoloader

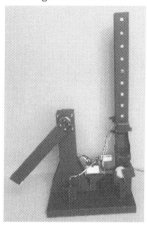

The throwing robot

The catapult robot

These robots shoot a ping-pong ball via three techniques: hitting, throwing, and catapulting. They shoot the ball up to a range of 6 m, which means that the ball reaches a launching speed up to 12 m/s. They are precision machines, which means that they are capable of pitching the ball into a wastepaper bin with a circular aperture size of 30 cm at a range of 6 m. The book gives you all the design drawings and plans to build the structure of the robots from 1.5 mm thick cardboard using glue and basic hand tools such as a box cutter, knife, rule, and pencil. All the step-by-step building instructions are given. You will need a few other materials such as aluminum tubing that is easily and cheaply available from hardware stores and there is only one specialized manufactured component, which is an aluminum flange for the stepper motor. Otherwise, you will need to purchase a Basic Stamp microcomputer board, a stepper motor and its driver, a servo, some sensors, and springs. You will learn how to build stiff, strong, and lightweight mechanical structures and moving mechanisms. You will learn the fundamental math and physics behind the analysis of these robots and you will learn how to program your machine that will bring the robots "to life" with real-time computer code that executes very interesting algorithms. What you learn has a strong generic component in that it will imbue you with the skills that will enable your technopreneurial and inventive talents. From these skills, you will be able to be a technopreneur by going it alone in the creation of your own working microcomputer-controlled prototype machines from scratch without expensive tools and equipment.

THE TEACHING PHILOSOPHY

Engineering is largely about researching, designing, and building products and processes that create, or make more efficient, a commercial business or other institution. As such, a large portion of engineering in society is tied up with business. Product design engineering requires careful thought, some risk-taking, knowledge of theory, experience of manufacturing techniques and methods and, last but not least, an analysis of safety issues and consequences when things fail. Cost and use of resources is also a major concern. We emulate many of these engineering features in this book, which details the theory and practice of creating precision robots. As such, engineering students can begin to understand what engineering is all about and at the same time gain valuable expertise in the creation of precision mechatronic machines and robots.

The book concentrates on mechatronics and robotics engineering and uses a low-cost technique using cardboard to build mechanical structures and mechanisms followed by adding sensors, servos, actuators, and microcomputers to bring the product to life. The authors see, in their experience, that there seems to be too much prevalence of report writing and imaginative design presentations and not enough making, doing, building, and operation of real working machines. Both are important but so is the balance.

The level of the book is designed for students and educators of engineering at, (i) university undergraduate level, (ii) upper level college students, or (iii) upper level preuniversity school students. The book is the product of teaching engineering undergraduate and postgraduate students in subject areas such as, "Mechatronics Project," "Innovation and Design," "Electronic Design Project," and Final Year Project/Capstone Year Project, plus numerous other engineering subjects over the past 25 years at the university level. There is, however, an application motive in developing the three-ball shooting robots in this book and it is one of devising a new game called "robot ping-pong basketball" that concerns mobile robots that can see, identify, and pick up a ping-pong ball and then throw the ball to a team member or shoot the ball into the opposing team's basket, that being a wastepaper basket. The idea is that the game will be developed over a period of years by students under supervision.

The book shows students how to use a "Glued-Cardboard Engineering" technique for the construction of robot mechanical structures and mechanisms. As such, the book has a very strong thread that is concerned with model making. Model making is usually the preserve of schools of architecture and product design but here we bring model making right to the core of engineering but with a difference which is the addition of electronics, motors, actuators, sensors, microcomputers, and real-time software. Glued cardboard is a highly effective, low cost, and versatile, rapid prototyping technique. A variety of procedures and concepts is illustrated in the book such as (i) shape, not material properties, is the dominant property that creates lightweight, stiff, and strong structures; (ii) mechanisms can be fabricated from cardboard; and (iii) three-dimensional shapes can be created from discrete elements. The authors have found that engineering students are very receptive to the intuitive understanding and absorption of theoretical and mathematical concepts if they can experience the workings of a real physical machine system or robot that they personally build and program with software code. The modern microcomputer is so small, cheap, and computationally powerful that fascinating machines can be built that work right in front of the student on the bench top. These machines can be statically bench-mounted but still have moving parts or can be mobile with wheels or legs or combinations of both. The machines can follow static equilibrium rules or can be dynamic such that they, for example, can jump or throw objects. Furthermore, students working in groups can build, in the duration of just one semester, their own working robot. The authors see students, some of whom have never built an artifact in their lives, gain a "can-do" confidence that ensures their growth through university so that they are more able in industry or in research when they graduate.

Throughout this book, the authors ensure that mathematical concepts and physical principles are rigorously described hand-in-hand with the design and constructional techniques of the working robot. This is of vital importance since such a methodology is the lifeblood of engineering. The reader is also constantly reminded of the importance of tolerancing and the correct use of numbers in programming because they relate directly to the cost and performance of the product being manufactured. Arguably, the only way of teaching the importance of tolerancing and the use of numbers is by practice in the building of working robots and so this book serves the student well in this respect.

The authors also see at present a highly welcomed movement in universities toward a more project-driven form of engineering education. The movement promotes a more inspiring focus that is on a learning by doing problem-based approach via working robots. One such indicator that supports our belief is that students become so engrossed in the design, construction, and programming of their robots, together with the problem-solving aspects of the teaching methodology, that they have to be shooed' out of the classroom even when time is up. You also find that students request extra time to do more work to improve further their robots because they develop a sense of ownership of the machine together with an obsessive interest in solving the given problems.

The authors believe that the product design engineering learning outcomes of this book can be applied to all science and engineering students at technical college, undergraduate, and post-graduate levels. Recently, the authors are realizing that a course based on this book would also be of interest to business students who would benefit from an understanding of engineering principles. Also, the learning outcomes have a wider application in the development of short courses for industrial professionals.

RESOURCES

Videos of the robot arms featured in this text can be viewed using the following links:

- Frank Nickols' website: www.franksrobomachines.com
- Mr. Robot showing the Hitter and the Catapulter: https://www.youtube.com/watch?v=WnxKX-q271Y

- Two students building the Catapult but adding their own radio-controlled omnidirectional mobility system: https://www.youtube.com/watch?v=HxEVSx5UFXA
- A student showing his trials and tribulations building the Thrower Robot: https://www.youtube.com/watch?v=2btPPKIpg8E&t=20s
- The Catapulter showing its repeatability competence: https://www.youtube.com/watch?v=ylJDFRihZcM
- Three clips of the Hitter
  - Hitter 1 https://www.youtube.com/watch?v=OyPv2J-IAV8
  - Hitter 2 https://www.youtube.com/watch?v=jPtY8S_mpZY
  - Hitter 3 https://www.youtube.com/watch?v=G_iYnFtFySc

Acknowledgments

We would like to thank Dr. Mani Le Vasan for her insightful comments, pedagogical guidance, and unstinting support throughout the writing of this book.

Chapter 1

Tools, Cutting Techniques, Risk, Reliability, and Safety Issues

LEARNING OUTCOMES

1. Ability to cut cardboard accurately by hand with a box cutter knife, potentially to ±0.2 mm error.
2. Awareness of, and accounting for, risk, reliability, and safety issues.

1.1 TOOLS REQUIRED FOR CUTTING AND GLUING CARDBOARD

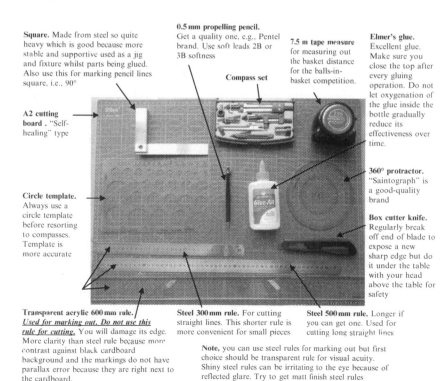

Square. Made from steel so quite heavy which is good because more stable and supportive used as a jig and fixture whilst parts being glued. Also use this for marking pencil lines square, i.e., 90°

0.5 mm propelling pencil. Get a quality one, e.g., Pentel brand. Use soft leads 2B or 3B softness

Compass set

7.5 m tape measure for measuring out the basket distance for the balls-in-basket competition.

Elmer's glue. Excellent glue. Make sure you close the top after every gluing operation. Do not let oxygenation of the glue inside the bottle gradually reduce its effectiveness over time.

A2 cutting board . "Self-healing" type

360° protractor. "Saintograph" is a good-quality brand

Circle template. Always use a circle template before resorting to compasses. Template is more accurate

Box cutter knife. Regularly break off end of blade to expose a new sharp edge but do it under the table with your head above the table for safety

Transparent acrylic 600 mm rule. *Used for marking out. Do not use this rule for cutting.* You will damage its edge. More clarity than steel rule because more contrast against black cardboard background and the markings do not have parallax error because they are right next to the cardboard.

Steel 300 mm rule. For cutting straight lines. This shorter rule is more convenient for small pieces

Steel 500 mm rule. Longer if you can get one. Used for cutting long straight lines

Note, you can use steel rules for marking out but first choice should be transparent rule for visual acuity. Shiny steel rules can be irritating to the eye because of reflected glare. Try to get matt finish steel rules

Creating Precision Robots. https://doi.org/10.1016/B978-0-12-815758-9.00001-8

1.2 **MARKING OUT THE CARDBOARD**

You need to order black sheets of A2 size and a cardboard of 1.5 mm thickness. This cardboard has an areal density of approximately $1.2\,kg/m^2$, which is useful when estimating the mass of moving parts such as the shuttle in the Catapult Robot.

Marking out should be done with the transparent rule as a first choice as there will be no parallax error because the rule markings are adjacent to the cardboard surface. Whether you use a transparent rule or a steel rule, which has parallax error due to its markings 1 mm above the cardboard surface, you should use the pencil as shown here to reduce any further possibility of parallax error.

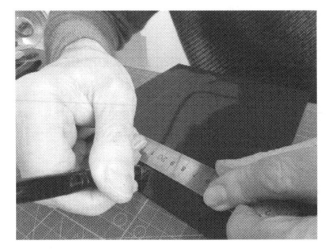

After making pencil marks you should draw pencil lines indicating where to cut the cardboard. Keep your pencil upright when viewed from the front as shown here. Remember the pencil lead is 0.5 mm but the metal spout that supports the lead has a diameter of 0.8 mm so if your lead is protruding greater than 1 mm which is the thickness of the rule then you are increasing the likelihood of error. So keep your lead short such that the metal spout is in contact with the rule and the rule is thus $0.8\,mm/2 = 0.4\,mm$ away from the pencil marks. It may seem impossible to obtain such precision but you will be surprised that with patience and practice, human eye and hand coordination is surprisingly accurate.

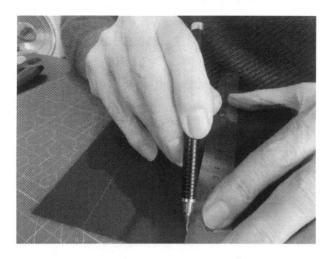

1.3 **CUTTING STRAIGHT LINES**

Once you have marked out straight lines the next step is to maintain your accuracy when cutting. What you do now is to place the tip of the knife right in the middle of your pencil line keeping the knife upright as shown in the figure. Resist the tendency to lean the knife out or in from upright.

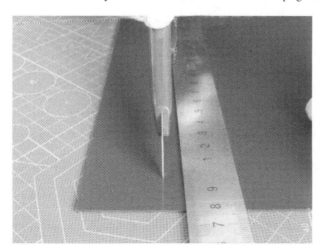

Keep your knife where it is then bring the steel rule up gently to touch the knife. Remember, do not cut using the plastic acrylic rule because you will damage its edge; that is only to be used for marking out.

Now, hold the rule where it is by pressing down with your forefinger (at the rule marking "10" in this picture) and do the same for the second pencil marking at the other end of the cardboard.

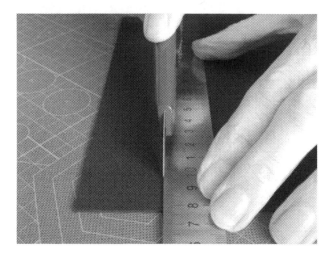

Now spread your fingers along and up and down the rule pressing down firmly and take great care to make sure that, with sufficient clearance, there is no flesh and bone in the line of the knife cut. Also make sure that there is nobody behind you who is likely to nudge you. Now take light cutting strokes to *take approximately eight cuts to cut through the 1.5 mm thick cardboard*. Remember to keep the knife upright as in the previous photograph but at the angle shown in the picture, about 20 degrees from the horizontal, when viewed from the side. If you cut at a higher angle you will tear the cardboard and if you cut at lower angle you are likely to steer the knife away from the rule or over the top of the rule. Keep the knife at about 20 degrees, and with the knife aimed a few degrees, to cut toward the rule thus the knife will follow accurately your pencil line. **But at the same time do not forget to keep the knife upright**.

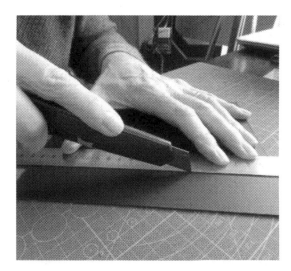

1.4 **CUTTING CIRCLES**

It is possible to cut accurate, very round circles if you follow these simple rules. Here you see a 16 mm-diameter circle being cut for the Catapult Robot. First, use the circle template to mark out a 16 mm diameter circle, making sure you keep the pencil upright. Now use the knife in a stabbing action with the normal 20 degrees cutting angle elevated to about 45 degrees. Now take many stabbing cuts around the pencil line, stabbing as accurately as you can through the center of the line. Rotate the cardboard ccw (counterclockwise) as you go and complete the circle in the cw (clockwise) direction. Your stabbing actions should take the tip of the blade completely through the cardboard into the self-healing cutting board.

Now do the same stabbing actions in the opposite direction. This time you do not have to press so hard. Complete the circle working in the ccw direction (cardboard rotated cw). The 16 mm diameter cardboard dime-sized center should be almost falling out. Do not press it down because it will tear on the back side. Instead push it up with your finger on the underside until it pops out. If it is stubborn you must repeat the stabbing cutting actions. Remember to keep that blade upright.

Here is the "dime" piece popped out. Cutting can get quite tedious and tiring especially when you have a lot of cutting to do. When you get tired then take a break and have a cup of tea. As students, you work as a team to make these robots, you should choose the student with the most patience and diligent care at making things to do the most difficult parts. Not everybody is cut out with these attributes.

1.5 BREAKING OFF OLD SECTION TO EXPOSE A NEW SHARP EDGE ON THE BOX CUTTER

After some time the cutting edge of the knife blade will lose its sharpness. What you do is to break off a section of the blade to create a new cutting edge. To do this, follow this procedure.

First retract the blade for safety before you remove the back part of the cutter

Note that the blade has a safety lock to inhibit the blade from sliding. Locking can stop the blade from slipping while cutting, which can save your fingers from being cut.

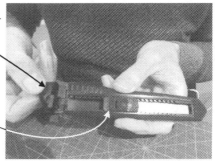

Second extend the blade by one section and insert the back part into the extended blade section then break it off under the table with your head above the table. If you do the break properly then the broken-off piece will not fly off which can be dangerous.

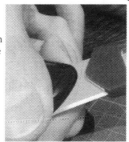

Third, doing it right means that the broken-off old blade section stays in the back part which is held between the thumb and the forefinger.

1.6 SAFETY CONCERNS FOR THE HITTING ROBOT

The major safety concern for the Hitting robot is the path of the arm extremity and the ball launching path which can cause personal harm. Students working together can inadvertently press the "hit ball" button without warning classmates to clear the area. So a safety check list and procedure has to be set up by students that may include wearing eye protection at certain times. Students have had to explain to their professor their risk assessment and safety check list and procedures and were encouraged to share ideas with competing team mates. Another potential hazard that concerns damage to the robot itself rather than human beings is the damage to property, that is, the ball tee.

This happened a few times because a ball released by the autoloader got wedged in the tee and the hitting arm attempted to drive through the ball and damaged the fragile tee. Other times the upper and lower gates connected to the gate servos were damaged because balls got jammed.

Thus there is a potential reliability problem as well as safety problem. This is a very real aspect of product design, that is, prevention of problems, where extra sensors are required to sense if a ball is there or not. Thus sensors should be considered to be placed in the windows of the autoloader and at the ball tee. Of course the solution does not end there because you have to write extra software code and integrate that into the ball hitting algorithm.

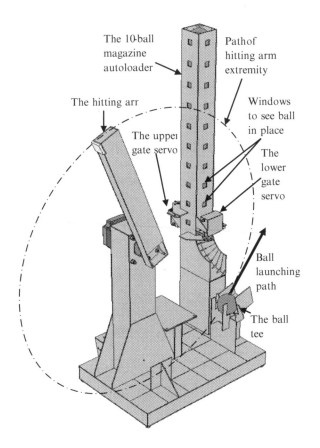

1.7 **SAFETY CONCERNS FOR THE THROWING ROBOT**

The major safety concern for the Throwing Robot is the same as the Hitting Robot, that is, the path of the whirling smart arm extremity causing personal injury. A similar safety procedure should be set up by students and that should be explained to the supervising professor. The smart arm tangential speed of its extremity reaches more than twice that of the hitting arm of the Hitting Robot. Thus extra care has to be taken with the smart arm safety.

Also the Honeywell HOA1405-2 light sensor can experience premature triggering due to ambient light entering the sensor and not just due to light being reflected from the white paper patch. This will cause the ping-pong ball fly out in unwanted directions. Students should consider an improved method for triggering the ball release servo such as replacing the Honeywell light sensor with a Hall-effect sensor, for example, AH337 and a magnet, that is, a small magnet replacing the white paper patch and the AH337 replacing the Honeywell sensor. Another method might be a magnetic pick-up sensor mounted on the arm and a small steel tooth replacing the white patch or a reed switch.

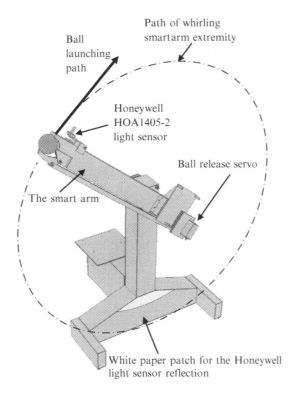

A significant learning component of this book concerns problem-based learning where students are presented with sufficient knowledge then they have to fly solo occasionally to improve the robot performance. Here is a prime example of such a learning procedure.

1.8 **SAFETY CONCERNS FOR THE CATAPULT ROBOT**

There are no safety concerns of a high-speed rotating arm with this robot. The potentially harmful moving components, that is, the shuttle and the springs that elevate the kinetic energy of the ball to a maximum speed of 12 m/s, are all contained within a safe boundary which is the two cardboard slideway tubes. After releasing the ball the shuttle and springs continue to move by overshooting safely inside the confines of the cardboard tubes until coming to rest after oscillating at the position shown in the picture. Thus probably the major safety concern is the ball launching path which is a concern for all the three robots. This means that students should devise a check list before initiating the catapult to clear students away from the ball trajectory. The ping-pong ball is an ideal projectile for these robots because table tennis has evolved over many years as a safe game because the ball weighs a mere 3 grammes-f and it is not required to wear safety glasses. If a European squash ball had been chosen or a golf ball then this would be very much a different situation.

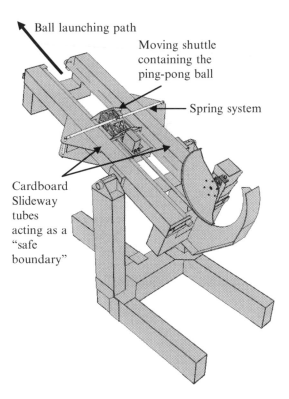

Ball launching path

Moving shuttle containing the ping-pong ball

Spring system

Cardboard Slideway tubes acting as a "safe boundary"

Theory I: Ball Trajectory Computation Using Excel Spreadsheet

LEARNING OUTCOMES

1. Derivation of a dynamical equation in equilibrium
2. Creating a discrete time-slice solution and determining its accuracy
3. Generating a graphical solution using Excel spreadsheet

2.1 INTRODUCTION

Inspiration for these hitting, throwing, and catapulting robots came from one of the author's experiences when working on the Buccaneer aircraft bomb delivery system in the 1970s. The Buccaneer was a fighter-bomber used by the British Royal Air Force. The pilot, flying the aircraft at 800 kph and at 16 m above the sea surface, so as to be below the radar horizon of the enemy target ship, Fig. 2.1, pulled the aircraft up at 7 degrees/s using a head-up display at 6 km range from a moving target ship and released a free-flight bomb. The trigger for the bomb release originates from an onboard analog computer which releases the bomb at a real-time computed angle from the horizontal that is dependent on (i) the aircraft ground speed, (ii) the radar measured range from the target, (iii) the wind velocity, (iv) the target velocity, and (v) the aircraft centripetal acceleration. Nowadays, bombs are non-free flight largely smart bombs but nonetheless free-flight principles are inherently fundamental to the trajectory of smart bombs.

It is similar, but more scientific, to the way we judge the angle and speed of throwing an apple core into a wastepaper bin, albeit a stationary bin. The mathematics of the bomb free-flight trajectory involves the solution of a nonlinear differential equation. This was difficult to solve at that time but nowadays it can be solved easily by using Excel spreadsheet and a discrete time-slice finite difference equation.

Creating Precision Robots. https://doi.org/10.1016/B978-0-12-815758-9.00002-X

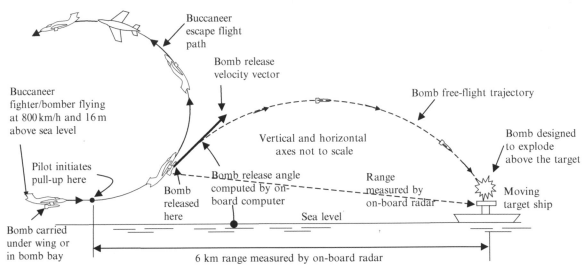

Buccaneer escape flight path

Bomb release velocity vector

Bomb free-flight trajectory

Buccaneer fighter/bomber flying at 800 km/h and 16 m above sea level

Vertical and horizontal axes not to scale

Bomb designed to explode above the target

Pilot initiates pull-up here

Bomb release angle computed by on-board computer

Range measured by on-board radar

Bomb released here

Moving target ship

Sea level

Bomb carried under wing or in bomb bay

6 km range measured by on-board radar

■ **FIG. 2.1** Inspiration for the student projects in this book came from author experience with weapon delivery systems.

2.2 **DISCRETE DYNAMICS EQUATION OF MOTION**

We now analyze the passage of a nonspinning, spherical ball through still air. The three forces acting on the ball, Fig. 2.2, during its passage through the air are as follows:

1. Aerodynamic drag force vector, $\underline{F}_{aero}$ which acts in the direction opposite to its instantaneous velocity vector, $\underline{v}$.
2. Gravitational force vector, that is, the weight, $\underline{W}$ which acts vertically downwards.
3. Inertial force, F_{in} which acts in the direction opposite to $\underline{F}_{aero} + \underline{W}$ such that the ball is in dynamic equilibrium, that is, sum of all static and dynamic forces is zero, $F_{in} = -(\underline{F}_{aero} + \underline{W})$.

Note that the ball acceleration vector is in the same direction as $\underline{F}_{aero} + \underline{W}$.

Now, let us analyze the aerodynamic drag force vector. Fig. 2.3 shows the aerodynamic force, $\underline{F}_{aero}$ acting on the ball that is travelling to the right at velocity, $\underline{v}$ through static air. The diagram illustrates the situation as if a camera is moving at the same velocity as the ball indicating that the ball is static so that it is easier to visualize the state of affairs.

The magnitude of the aerodynamic drag force vector $\underline{F}_{aero}$ is,

$$\mathbf{F_{aero}} = \tfrac{1}{2} \cdot C_D \cdot \rho \cdot A \cdot v^2 \tag{2.1}$$

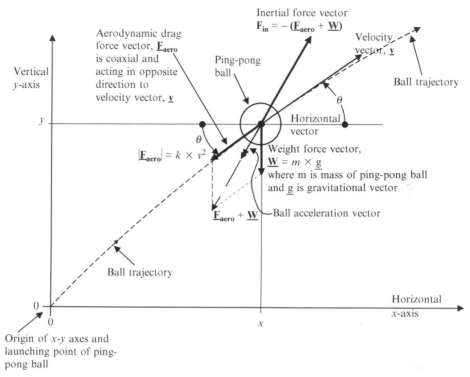

■ **FIG. 2.2** Instantaneous vector diagram of forces, velocity, and acceleration of ball during its trajectory in a vertical *x-y* plane.

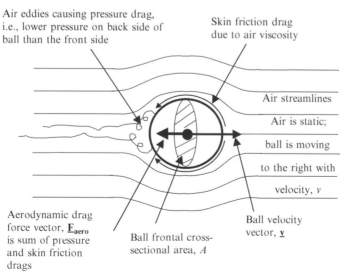

■ **FIG. 2.3** Diagram showing drag force vector, $\underline{F}_{aero}$ acting on ping-pong ball travelling through air at velocity $\underline{v}$.

where

C_D = coefficient of drag (dimensionless)
ρ = density of air
A = cross-sectional area of ball orthogonal to ball velocity vector
v = speed of ball through the air

Note that C_D, ρ, and A will be considered constant values. The coefficient of drag, C_D is considered a constant value because it will be assumed that the Reynolds number is high enough to cause turbulent flow.

Thus the magnitude of the ball aerodynamic acceleration vector, $\underline{\mathbf{a}}_{\mathbf{aero}}$ due to aerodynamic force is, from Newton's 2nd law ($F = m \times a$, hence, $a = F/m$), given by

$$|\mathbf{a_{aero}}| = \frac{\mathbf{F_{aero}}}{m_{ball}} = \frac{^1/_2 \cdot C_D \cdot \rho \cdot A \cdot v^2}{m_{ball}} = k \cdot v^2 \tag{2.2}$$

where

m_{ball} = mass of ping-pong ball
k, named the "deceleration drag constant," is an important dimensioned constant that is given by

$$\text{Deceleration drag constant, } k = \frac{^1/_2 \cdot C_D \cdot \rho \cdot A \; (m^{-1})}{m_{ball}} \tag{2.3}$$

Note that the deceleration drag constant, k has unit m^{-1}. Be careful to note that the direction of aerodynamic acceleration, $\underline{\mathbf{a}}_{\mathbf{aero}}$ is opposite to that of the velocity vector, $\underline{\mathbf{v}}$. The vector $\underline{\mathbf{a}}_{\mathbf{aero}}$ is now separated, conveniently, into two orthogonal directions, that is, in the horizontal x and vertical y directions, Fig. 2.4. These two forces are used to derive two of the three accelerations (the third is gravitational acceleration), acting on the ping-pong ball that are assumed to act at constant values for the duration of time Δt from t_n to t_{n+1}.

The three accelerations are as follows:

1. Gravitational acceleration, $g = -9.81 \; (m/s^2)$ that acts vertically.

Negative sign indicate a direction opposite to the positive y direction.

2. The horizontal acceleration component of $\underline{\mathbf{a}}_{\mathbf{aero}}$, which is:
   $-k v_n^2 \cos \theta_n \; (m/s^2)$

Negative sign indicate a direction opposite to the positive x direction.

3. The vertical acceleration component of $\underline{\mathbf{a}}_{\mathbf{aero}}$, which is:
   $-k v_n^2 \sin \theta_n \; (m/s^2)$

Negative sign indicate a direction opposite to the positive y direction.

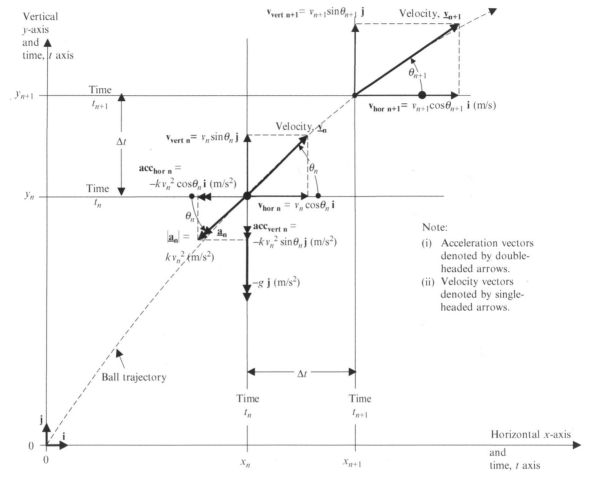

■ **FIG. 2.4** Setting up the ping-pong ball discrete time-slice mathematical model trajectory using Excel spreadsheet.

So total vertical acceleration component is, $acc_{vert} = -\left(g + kv_n^2 \sin\theta_n\right) \, (\text{m/s}^2)$.

These three accelerations are used to determine the next value of velocity vector $\underline{v}_{n+1}$ using the classic equation for constant acceleration, that is, $v = u + at$ as follows:

1. Horizontal component of $\underline{v}_{n+1}$ is given by

$$v_{n+1 \, hor} = v_n \cos\theta_n - kv_n^2 \cos\theta_n \times \Delta t$$

2. Vertical component of $\underline{v}_{n+1}$ is given by

$$v_{n+1 \, vert} = v_n \sin\theta_n - \left(kv_n^2 \sin\theta_n + g\right) \times \Delta t$$

Thus the new velocity vector, $\mathbf{v_{n+1}}$ is given by

$$\mathbf{v_{n+1}} = \sqrt{\left(v_{n+1\,hor^2} + v_{n+1\,vert^2}\right)/\mathrm{atan}\left(v_{n+1\,vert}/v_{n+1\,hor}\right)}$$

Meaning that: $\theta_{n+1} = \mathrm{atan}\left\{\dfrac{v_{n+1\,vert}}{v_{n+1\,hor}}\right\}$

and the magnitude of $\mathbf{v_{n+1}}$ is $\sqrt{\left(v_{n+1\,hor^2} + v_{n+1\,vert^2}\right)}$

The new x and y positions at t_{n+1} are given by

$$x_{n+1} = v_n \cos\theta_n \times \Delta t$$

$$y_{n+1} = v_n \sin\theta_n \times \Delta t$$

and the new x and y positions are computed as

$$\begin{aligned} x_{n+1} &= x_n + v_{\mathrm{hor}\,n} \cdot \Delta t \\ y_{n+1} &= y_n + v_{\mathrm{vert}\,n} \cdot \Delta t \end{aligned} \tag{2.4}$$

The calculation algorithm is then repeated to produce the velocity and x, y positions at t_{n+2} and so on iteratively that leads to the ball trajectory in x, y space. Before the Excel spreadsheet algorithm is described we need to calculate the ping-pong ball acceleration drag coefficient, k.

2.3 CALCULATING THE PING-PONG BALL ACCELERATION DRAG COEFFICIENT, k

In order to carry out the Excel spreadsheet algorithm with actual numbers, students are asked to search the internet for values of C_D for a spherical object together with the mass and diameter of a ping-pong ball and the density of air in order to calculate an estimate for the value of k. The value is considered an estimate because it will be evaluated approximately from published data then calibrated based on our experimental results when using the robot shooters. The results are as follows:

1. C_D for a sphere $= 0.47$ (estimated from Fig. 2.5 and its notes)
2. ping-pong ball diameter $= 0.040$ m (by measurement)
3. ping-pong ball mass $= 0.0027$ kg (by measurement)
4. density of air at $20°C = 1.21$ kg/m^3 (from internet)

Hence the value of the deceleration drag constant, k for the ping-pong ball is obtained by substituting values into Eq. (2.3) as follows:

$$\text{Deceleration Drag Constant,} \quad k = \frac{1/2 \times 0.47 \times 1.21 \times \pi \times 0.040^2}{0.0027 \times 4} \tag{2.5}$$

Note that k has been shown to a precision of three significant figures which means that the value has an error of $\pm 0.0005/0.146 \times 100\% = \pm 0.3\%$.

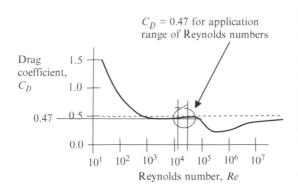

Notes:
Reynolds number, *Re*, for the ping-pong ball is given by the following reasoning.

$$Re = \frac{\rho v d}{\eta}$$

where;

η/ρ = air kinematic viscosity=15E–6 (m²/s) at 20°C
v = velocity of ball which is 4.7 m/s lowest (estimated) for the 2-m range target and 11.1 m/s (estimated) for the 6-m range target.
d = diameter of ball = 0.038 m

Thus lowest Re_{low} = $\frac{4.7 \times 0.038}{15E–6}$ = 1.2E4

and highest Re_{high} = $\frac{11.1 \times 0.038}{15E–6}$ = 2.8E4

From the NASA graph it can be seen that this range of *Re* numbers falls in an area where the drag coefficient, C_D remains steady at 0.47

■ **FIG. 2.5** Drag coefficient for a sphere versus Reynolds number. *(Source: NASA.)*

This error implies that *k* is rather better than an estimate but nonetheless students should correct the value later. Interestingly, the deceleration drag constant can be calculated for many other balls and objects that are thrown, kicked, or hit, for example, cricket ball, badminton shuttlecock, football, basketball to name a few. By entering just this one value, *k*, an Excel spreadsheet can be used to predict the free flight of most nonspinning balls or objects.

Students are now in a position to implement an algorithm using Excel spreadsheet to calculate the ball trajectory given an initial launch velocity.

2.4 EXCEL SPREADSHEET TABULATION, COMPUTATION, AND GRAPHICAL RESULTS

In order to calculate the trajectory of the ball in the *x*-*y* plane as *f(x,y)*, the ball acceleration components in the *x* and *y* directions are derived from the only two forces considered to be acting on the ball, that is, aerodynamic force, $\mathbf{F_{aero}}$ and the weight, $\underline{\mathbf{W}}$. Fig. 2.6 shows the layout for an Excel spreadsheet computation of the ball trajectory. A time slice of 1ms and 20,000 computation intervals are used thus allowing for a flight of 20 s. Note that the shaded cells are values that are entered as variables which are (i) the discrete time-slice value, Δt, (ii) the value of the deceleration drag constant, *k*, and (iii) the two values for the ball launch velocity at time, *t*=0, that is, v_0 and θ_0.

Fig. 2.7 shows a graphed result of the computation with $\Delta t = 0.001$ s, $k = 0.146 \text{m}^{-1}$, $v_0 = 10$ m/s, and $\theta_0 = 40$ degrees. On the same graph, for comparison, is shown the ball trajectory with zero air resistance, that is, $k = 0 \text{m}^{-1}$.

Time slice calcu-lation number n	Duration of discrete time slice Δt (s)	Time at start of time slice t_n (s)	Decel-eration drag constant k (m^{-1})	These values all at start of time slice							
				Ball launch velocity		Horizontal component of velocity $v_{hor\,n}$ (m/s)	Vertical component of velocity $v_{vert\,n}$ (m/s)	Horizontal acceleration component $acc_{hor\,n}$ (m/s^2)	Vertical acceleration component $acc_{vert\,n}$ (m/s^2)	Ball horizontal position x_n (m)	Ball vertical position y_n (m)
				Ball speed v_n (m/s)	Ball angle θ_n (°)						
0	0.001	0.000	0.146	10.00	40.00	7.66	6.43	11.18	9.38	0.0000	0.0000
1	0.001	0.001	0.146	9.98	39.96	7.65	6.41	11.14	9.34	0.0077	0.0064
2	0.001	0.002	0.146	9.96	39.91	7.64	6.39	11.11	9.29	0.0153	0.0128
3	0.001	0.003	0.146	9.94	39.87	7.63	6.37	11.07	9.24	0.0229	0.0192
⋮	⋮	⋮	⋮	⋮	⋮	⋮	⋮	⋮	⋮	⋮	⋮
1174	0.001	1.174	0.146	6.00	−60.29	2.98	−5.21	2.50	−4.38	5.5631	−0.4914
1175	0.001	1.175	0.146	6.01	−60.34	2.97	−5.22	2.50	−4.39	5.5661	−0.4966
1176	0.001	1.176	0.146	6.01	−60.38	2.97	−5.22	2.50	−4.40	5.5691	−0.5018
1177	0.001	1.177	0.146	6.01	−60.43	2.97	−5.23	2.50	−4.40	5.5721	−0.5071

■ **FIG. 2.6** Excel spreadsheet table of values that shows computation of ball trajectory with a given initial launch velocity and air drag deceleration constant, k.

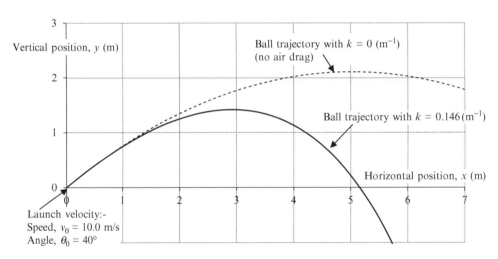

■ **FIG. 2.7** Graph of ball trajectory computed by Excel spreadsheet showing two trajectories; one with a realistic aerodynamic drag and the other with zero air drag to compare the effect that air drag has on the ball.

2.5 NOTES ON FIRST, SECOND, AND SUBSEQUENT ROW CALCULATIONS AND VALUES

Fig. 2.8 shows the calculations and values for the first row of trajectory computation at time, $t_0 = 0$. Fig. 2.9 shows the calculations and values for the second row of computation.

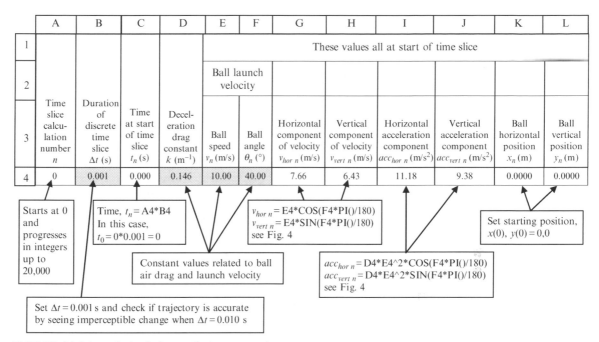

FIG. 2.8 Calculations and values for first row of trajectory computation.

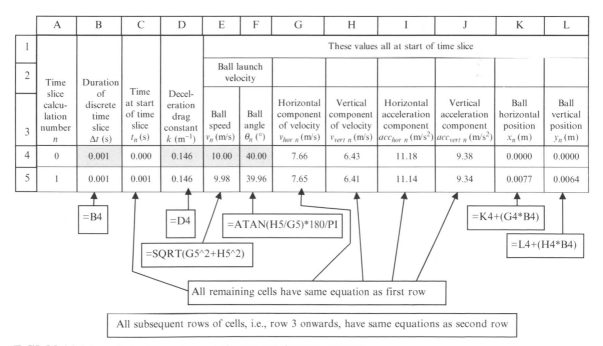

FIG. 2.9 Calculations and values for second row and subsequent rows of trajectory computation.

2.6 **USING THE TRAJECTORY COMPUTATION TO SYNTHESIZE SOME USEFUL EQUATIONS**

Now let us use the Excel spreadsheet to deduce some interesting methods for pocketing the ping-pong ball into the target wastepaper basket set at a height of 1 m and at a range of 5 m, Fig. 2.10. The radius of the basket mouth aperture is 0.15 m so let us set the permissible horizontal error of the ball in the x-y plane of the trajectory as it enters the wastepaper basket to be ±0.05 m. The center point of the basket mouth is to be named the target point, Fig. 2.10. The trajectory spreadsheet will now be used to synthesize an empirical equation for the ball launch velocity given the range of the basket. However, the situation is complicated by there being many solutions.

For example, if the launch speed is higher than necessary, that is, the energy expended on the ball is excessive, then Fig. 2.10 shows two solutions, launch angle = 35 and 55 degrees, for pocketing the ball at 5 m range. It turns out that using a trial-and-error approach the minimum launch speed for the two launch angles to converge to two equal launch angles to hit the target point is 11.06 m/s and the two equal launch angles are 45 and 45 degrees. Thus, there will be an infinite number of solutions to the launch angle for launch speeds greater than 11.06 m/s.

So, there needs to be a strategy for setting the launch velocity and one such strategy is as follows.

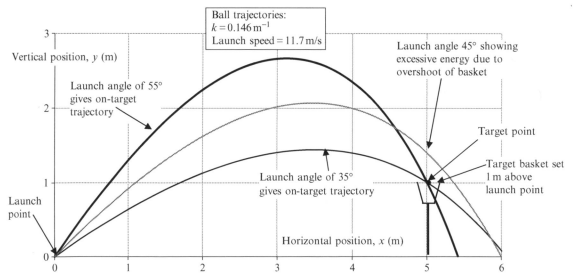

■ FIG. 2.10 Target set at 1metre height and 5 m range. Computed ball trajectories for three launch angles showing there are two solutions for launch angle for an on-target trajectory if the ball launch speed is excessive.

2.7 STRATEGY FOR SETTING LAUNCH VELOCITY USING THE EXCEL SPREADSHEET TRAJECTORY COMPUTATION

The ball target range is 5.00 m with a tolerance of ±0.05 m. The strategy for setting a unique launch velocity as a function of target range, given that the target is 1 m above the launch point, as shown in Fig. 2.10, will be formulated by setting a minimum launch speed, that is, minimum ball energy, that will reach the target as well as the maximum permissible error of +0.05 m, that is, a target range of 5.05 m. We will assume a deceleration constant of $k = 0.146\,\mathrm{m}^{-1}$. Minimum ball energy means that the two launch angles should merge to give equal values at this range. The idea behind this strategy is that following the formulation of the minimum ball speed, the launch angle variation above and below is computed such that the ball hits the target at its lowest tolerance of -0.05 m, that is, a target range of 4.95 m. Using the Excel spreadsheet trajectory computation, Fig. 2.11 shows a range of launch angles and launch speeds that cover the target range from 4.95 to 5.05 m. From the graph it can be seen that if we set the launch speed range from 11.06 to 11.14 m/s and the launch angle range from 41.4 to 49.0 degrees then

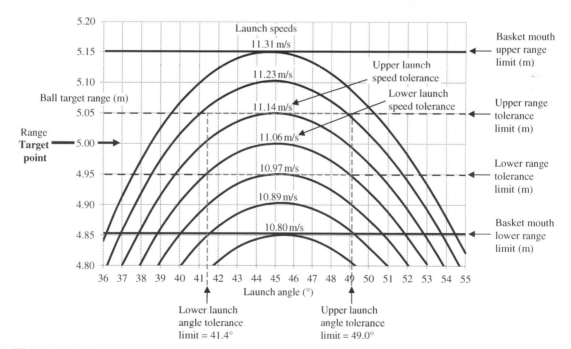

■ **FIG. 2.11** Graph of ball range with respect to target for various launch speeds and launch angles. Graph also shows that if the launch speed is 11.10 ± 0.04 m/s and the launch angle is 45.2 ± 3.8 degrees then the ball range will be 5.00 ± 0.05 m.

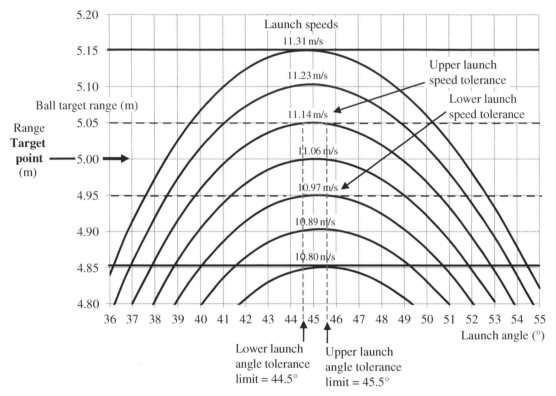

■ **FIG. 2.12** Graph showing how widening the launch speed tolerance narrows the launch angle tolerance.

the ball will achieve a target range of 5.00 ± 0.05 m. This is not a unique solution. For example, if the launch speed range is widened from 10.97 to 11.14 m/s then the launch angle range is narrowed to 45 ± 0.5 degrees, see Fig. 2.12. Hence, if the robot can achieve high launch angle precision then there is less demand for precision for launch speed and vice versa.

2.8 COMPETITION SCORING SYSTEM FOR POCKETING BALLS IN BASKETS

Students are taught how to build a hitting, throwing, or catapulting robot or a robot of their own ingenuity over the duration of 12-week semester. In the final week the students compete in a "Balls-in-Basket" competition. The competition gives the students 80% of their final marks. The remaining 20% marks is from a two-clip video report that shows the construction of their robot and what they have learnt from building and operating it. The first clip is a 30 s engineering skill aptitude selling pitch, or "elevator pitch" that the students can show at a job interview and the second clip shows more technical detail in no more than 2 min. These clips can be archived as a chapter in their work portfolio, similar to architecture and product design students.

Rules of the Ball-in-Basket Competition

Here are the rules of hitting, throwing, or catapulting (projecting) balls into basket. The basket is a metal mesh wastepaper bin with a 30 cm circular aperture that has a nice "clang" sound when the ball drops into the basket. The clang is good to hear but sometimes the ball bounces out of the basket, so to stop this, place a plastic bag or bubble wrap at the bottom of the baskets to reduce the coefficient of restitution to near zero. Many balls will hit the rim of the basket and bounce off so do not let students wrap plastic around the rim of the basket. This is cheating.

1. Balls are to be projected into a target basket set at 2 m range. If a ball lands in the basket and it stays there, that is, not bounce out, then the ball is "pocketed." The range is measured from the center of the tower of the hitting and throwing robot to the center of the basket. You can have the basket set at different levels with respect to the robot, for example, robot and basket both on the floor, table, or the robot on the floor and the basket set on a stool.

2. You have 10 ping-pong balls and cannot reuse the same ball and balls should be checked for tampering, for example, adding mass which will increase accuracy. Why is that? Students should explain this with Eq. (2.3).

3. As soon as three balls are pocketed then the game is finished, that is, no need to project any more balls.

4. The scoring system is as shown in Fig. 2.13. For example, if you pocket the first three balls of your 10 balls then you get maximum marks of $3 \times 10 = 30$ marks. If you miss the target then you keep projecting until three balls are pocketed which is when you stop, or your 10 balls are used up because you have to stop. An example of a low score might be that you pocket the last three balls of your 10 balls. In this case you get a score of $5 + 4 + 3 = 12$ marks. If you do not pocket any of your 10 balls then you get zero marks. Or, for example, if your 10th ball is pocketed then you score 3 marks and no more chances, the game is over.

5. The competition involves three games so the projection is repeated at 4 m range, and then at 6 m range. The competition becomes more difficult at greater ranges. Most student teams get full marks for the 2 m range and success at 6 m range. Students should explain why the error

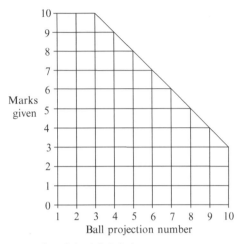

increases with range. In so doing, students buildup their scientific rigor and then realize how to make their machine more accurate.

2.9 CONFIRMATION OF THE DECELERATION DRAG CONSTANT, k (m^{-1})

The "aerodynamic deceleration constant," k will now be renamed to more simply, the "drag constant," k. If the value of k is known accurately, and given that we know the range and height of the target basket, the Excel spreadsheet model will be very useful in calculating the ball launch speed and angle required to deposit the ball in to the wastepaper basket. Here is a suggested method for confirming the value of k.

First let it be assumed that a simplifying assumption for confirming or estimating the value of drag constant, k, is to assume that k is a function of (i) the trajectory maximum flight height, h_p, Fig. 2.14, (ii) the horizontal distance, L_p, where the maximum height occurs, (iii) the difference in height between the ball launch point and the landing point, L_H, and (iv) the horizontal distance of the maximum flight distance, L_R. Measuring L_R is straightforward if a wrinkled carpet of plastic is laid where the ball strikes the floor. In this case the ball will not bounce so it is easy to measure L_{max}. However, when doing the experiment, make sure that you measure the difference in height between the launching point of the ball and the point of hitting the floor. Measuring L_p, and h_p, is more difficult and a measuring stick is suggested together with the help of a classmate.

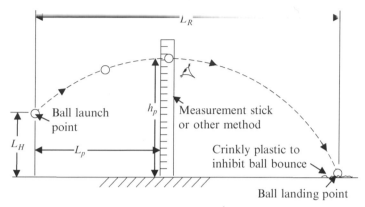

■ **FIG. 2.14** Experiment to confirm the drag constant, k (m^{-1}).

Here is a suggested method to follow in order to measure the value of k.

1. For the throwing robot, measure the angular velocity, ω of the arm at ball launch using an oscilloscope.
2. Convert the angular velocity to launch speed, $v = \omega r$.
3. Carry out experiments at the same launch speed but varying the launch angle in order to achieve the maximum flight distance.
4. Assume that the maximum flight distance L_R occurs at a launch angle of 40 degrees.
5. Use Excel spreadsheet with the ball launch velocity set to v, and the launch angle set at 40 degrees. Adjust the spreadsheet value of k, starting at the nominal value of $0.14 \, \text{m}^{-1}$ such that the spreadsheet maximum distance matches the measured value of L_R.
6. Check the measured values of L_p and h_p against spreadsheet values and if there is a difference then the assumed launch angle of 40 degrees may be wrong. In this case, adjust the spreadsheet launch angle and the value of k to match the measured three values of L_R, L_p, and h_p.

A much better idea than the measuring stick is to identify a smart phone app that can take a movie of the ball in flight that shows each frame of the movie so that the ball trajectory can be plotted. Once you think that the Excel spreadsheet is accurate then your spreadsheet can be used as a computing engine for setting the values of launch velocity in order to model a given trajectory. Your robot can now become an extremely interesting machine that, given the values of L_R and L_H, can hit, throw, or catapult the ball to a target. When your machine can (i) measure L_R and L_H autonomously, (ii) set the launch velocity autonomously, and (iii) aim the ball in the direction of the target basket autonomously, then you can say, arguably, that your machine is an intelligent agent, that is, it possesses computational intelligence and borders on possessing artificial intelligence.

2.10 **PROBLEMS**

1. Carry out hitting tests to confirm the coefficient of restitution of the ball-hitter arm interface by measuring the initial and final speeds of the ball and hitter arm using a smart phone.
2. Investigate if the ball is impeded when released from the gripper in the Throwing robot.
3. Investigate if the ball is given an initial spin with the Throwing robot, maybe, by marking the ball with a black pen and using a high frame rate camera.
4. Using the Excel spreadsheet model, investigate the launch angle of a golf ball for maximum distance. Golfers seem to hit the ball at a much lower angle than 40 degrees, possibly 20 degrees but the ball may have back spin which produces the Magnus effect.
5. Investigate the engineering feasibility of giving launch spin to the ball when using either the hitter or the thrower. For example, can the angle of the hitter plate be modified to give a David Beckham curve ball shot?
6. Fig. 2.7 shows a ball trajectory that was computed with a 1 ms time slice. At what value does the time-slice interval become inaccurate? What is defined as inaccurate when we do not have an exact solution to compare against?

3

Theory II: The Basic Stamp Microcomputer

LEARNING OUTCOMES

1. Knowledge of the digital binary input and output internal electronic switching methodology of microcomputers.
2. Software programming an analog-to-digital converter (ADC) using microcomputer input-output port switching together with a resistor-capacitor (RC) circuit.
3. Understanding techniques for precision computing with integers.

3.1 INTRODUCTION TO THE STAMP MICROCOMPUTER

The Basic Stamp microcomputer modules are designed and manufactured by Parallax. The modules are of the size of a postage stamp; hence its name and all the robots in this book are controlled by a Basic Stamp 2 Board of Education USB version, Fig. 3.1.

The first module, the Basic Stamp1, entered the market in the early 1990s. Now the modules have progressed to the up rated "2" versions, that is, BS2, BS2sx, BS2px, and others. The Parallax Company is highly dedicated to the education sector and the Parallax website, www.parallax.com hosts many educational tutorials and much information on understanding the instruction set. As such this chapter will not repeat the information that is available on the Parallax website. Instead what will be done is to give information related to student knowledge shortfall, based on the authors' teaching experience. This shortfall relates to digital binary input/output pins especially, the precision computation with 16-bit integers and the use of the ingenious RCtime instruction that is a very clever algorithm and method that measures the value of a resistance or capacitance. The value of this resistance or capacitance can represent a sensor value such as a knob angle of a potentiometer wired as a variable resistance or the distance between two plates that form a capacitance. This means that the RCtime instruction can be used as an analogue-to-digital converter, ADC.

Students often ask how this instruction works and it turns out that the instruction is worthy of a 2 h lecture and practical class where knowledge

Creating Precision Robots. https://doi.org/10.1016/B978-0-12-815758-9.00003-1

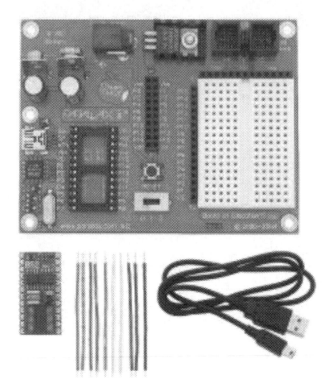

■ **FIG. 3.1** The Basic Stamp2 BS2, Board of Education USB version. This microcomputer board is used to control all the robots in this book.

of analog electric circuit theory, digital electronics, and low-level micro-computer programming all come together. Incidentally, "low level" pro-gramming is more difficult to understand because it operates at a more detailed level. Also, the RCtime instruction requires the use of the oscillo-scope to see the timing of the capacitor being charged and discharged and how the charge rate is controlled by the value of the resistor. Practising the use of the oscilloscope, which is one of the most valuable tools in electron-ics, coupled with the learning outcomes of the RCtime instruction mean that students experience an effective investigative and learning opportunity.

First of all it is necessary to understand how microcomputers in general, not just the Basic Stamp microcomputer, are wired internally to produce input and output ports or pins. Strictly speaking, a port consists of a group of 8 pins, which is historical due to 8-bit microprocessors, but that is subject to variation meaning that a port can be of 4 pins, 8 pins, 16 pins, or other numbers. The input/output, i/o, pins form a method of communicating with, and controlling of, the outside world. If there were no i/o pins then that would be like a brain without its five senses (input pins) and without the ability to control muscles (output) to move its body. The term "input" means that a pin is set to measure a digital voltage which means that if the voltage on the pin is below 1.5 V then it is registered as a logic low/logic 0 and if the voltage is above 3.5 V then it is registered as a logic high/logic 1. These voltage thresholds vary depending on the manufacturing technology and if the voltage on the pin is between 1.5 and 3.5 V then the logic state is indeterminate and depends on the previous state registered. Digital logic that works at these voltage levels is generally defined as "TTL logic" where TTL means "transistor-transistor-logic." The TTL was an early digital electronic manufacturing technology invented in the 1960s, which was quite suitable for the development of digital computers. The input side of a TTL logic gate has a very high input impedance which means that when a voltage is applied to its input, let's say a logic high of 5 V, then less than a few microamps will flow into the gate meaning that the input imped-ance is of the order of Megohms which is very high. Conversely, on the output side of a TTL gate, the output impedance is quite low, of the order of 25 Ω as measured by the author and is shown in Fig. 3.2. Well, are these values good or bad and of what consequence are they? Well, they are good because logic gates consume very little current at their inputs means that the logic gate that sent them a logic value provided very little current. What is the significance of this? Well, any current provided by a gate has to pass through the 25 Ω output impedance and thus there will be an associated voltage drop across it and thus a logic high of 5 V will be compromised by its value dropping to a lower value. So to conclude, logic gates should have high input impedance and low output impedance.

The ingenious thing about i/o pins is that the same pin can be reset as an "output" pin. In this case, that pin can no longer measure a digital voltage but this time it can output a digital voltage and that voltage can do some seri-ous controlling of actuators, motors, and even sensors. If you measure the voltage on that pin with a voltmeter then you will measure 0 or 5 V depending on the logic state demanded by the microcomputer program, that is, a logic low (0) or a logic high (1), respectively. Now the real ingenious thing about all these is that you can program to read the state of a pin when the pin has been programed as an input pin or you can program the pin to be a 0 V signal or a 5 V signal if the pin has been programed as an output pin.

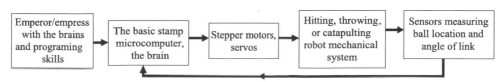

■ **FIG. 3.2** The fun of creating mechatronics and robotics systems.

Thus, here is the real power for the Mechatronics and Robotics engineer, that is, you are able to control a mechanical system with electric motors or other actuators and able to read sensors attached to the robot that tell you its position, velocity, or acceleration, for example, and then control its motion; and all this under fingertip control at the keyboard. It's a magical moment after spending a lot of effort and patience building your mechanical and actuator system to bring your creation to life at the keyboard with real-time computer code. You are like a modern-day Emperor or Empress with power at your fingertips and imagination, Fig. 3.2. We hope, too, that many other students feel inspired with the same reaction after building the robots in this book.

Ok, let's get started with explaining how one single i/o pin works.

3.2 **THE STAMP MICROCOMPUTER INPUT/OUTPUT (I/O) PORTS AND PINS**

Fig. 3.3 shows the inside of a Basic Stamp microcomputer where we use pin 0, that is, P0, as an example. Remember that almost all computers work in this way. The Stamp has 16 i/o pins and each one works completely independently.

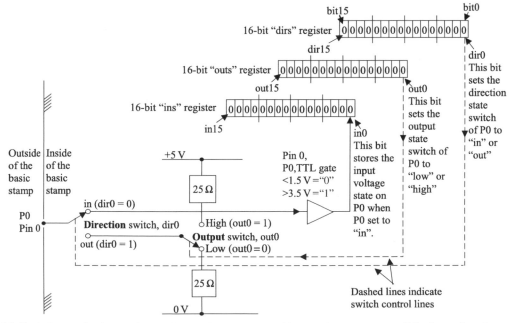

■ **FIG. 3.3** Electric circuit switch and register diagram of the Basic Stamp i/o, programmable circuitry for just one input pin, P0 The diagram shows the principle of i/o operation. Here, P0 is set to be an input on the command of bit0 of the dirs register.

In other words, you can have any combination of the pins being set to input or output. There are three 16-bit registers that individually manage the 16 i/o pins via the individual bits, for example, bit0, bit1, and bit15 of each of these registers. These registers are called the DIRS register, the INS register, and the OUTS register which are all under read and write program control of the Parallax Basic programming language, called PBasic. They are, in fact, 16-bit read/write memory registers that are volatile meaning that they can only hold their memory values while power is applied. They also default to all bits being set to zero on power up. The DIRS and OUTS registers control the status of electrical switches that are transistor switches. These switches are shown in Fig. 3.3 as mechanical switches which are analogous to transistor switches. The INS register is different because it does not control the status of switches but instead stores the digital voltage status of each of the individual input pins from P0 to P15. If, however, any one of the pins is not set to input then there will be no voltage read by that pin and any state held by that register bit will remain unchanged. You can use upper case or lower case to identify registers so, dirs, ins, and outs are ok to use. The "s" at the end of each of the registers means the plural of each register meaning that, for example, the DIRS register controls not 1 but all the 16 input pins, whereas DIR0 and OUT0 control only P0, meaning pin 0. Similarly, dir15 controls the direction switch of P15 via bit15 being set to 1 (P15 is out), or reset to 0 via bit0 being reset to 0 (P15 is in). Likewise, IN0 (bit0 of the ins register) reads only P0 for its digital voltage status as long as DIR0=0 (direction register bit0=0). Similarly, IN15 (bit15 of ins register) reads P15 digital voltage status as long as DIR15=0 (dirs register b15=0). As such, for any pins set to output, their "in" bits can be used as extra memory.

On the point of memory, the Basic Stamp2 microcomputers have only sixteen 16-bit words for its volatile RAM memory. This is a very small RAM memory but is not a limitation if you create very efficient algorithms. In fact, the limitations of the Stamp microcomputer in terms of 16-bit integer-only calculations and small memory size, both volatile RAM memory and non-volatile code memory, necessitate computational invention and efficiency. Of the 16 words available as RAM, the 16-bit dirs register, the 16-bit ins register, and the 16-bit outs register consume 3 of the 16 words available for RAM. Thus any "in" bits that can be used for RAM memory because their pins have been set to "out" can be very useful and represents a technique for squeezing more memory from the Stamp.

Fig. 3.3 shows the i/o circuitry programed to make P0 as an output pin. This is done by setting bit0 of the DIRS register to "1" which in turn (follow the dashed line from bit0) changes the DIR0 direction switch to the OUT direction. Thus P0 is now connected to the output switch, OUT0. If now, OUT0 is switched to ground by setting bit0 of the OUTS register to "0," that is,

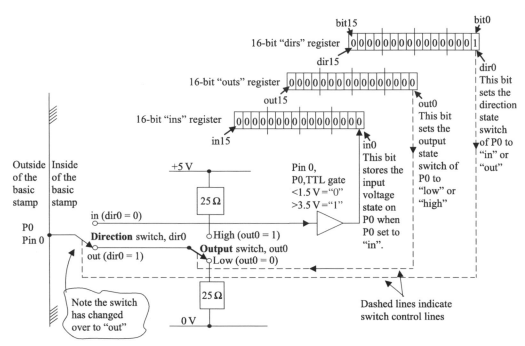

■ FIG. 3.4 Here, P0 is set to be an output on the command of bit0 = 1 of the dirs register.

OUT0 = 0 then this action will connect P0 to 0 V, meaning logic low. If, on the other hand, OUT0 = 1 then the OUT0 switch will be switched to the +5 V rail and thus a logic high will be sent to P0.

There are 15 identical copies of the circuit shown in Figs. 3.3 and 3.4. Each copy is dedicated to each of the remaining i/o pins, P1 to P15. Fig. 3.5 shows three identical circuits for P0, P1, and P15. There is not enough space and it would be very confusing to show all the 16 i/o circuits. You will notice that each bit of the DIRS, INS, and OUTS registers is dedicated to the relevant pin number. The INS register bits are all marked with "X" which means "unknown" or sometimes in other applications "don't care."

Ok, students may still be confused with the description of computer i/o workings so let's do some experiments to clarify.

3.3 EXPERIMENT #1. READING A LOW AND A HIGH ON AN INPUT PIN

This experiment, Fig. 3.6, connects a 10 kΩ potentiometer between the regulated +5 V power supply and 0 V where the pot wiper is connected to P0. The +5 V regulated power supply is available on pin 21 on the Stamp chip

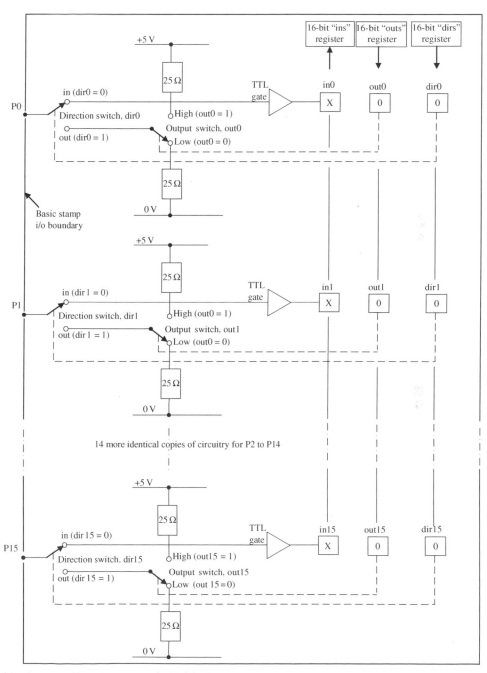

■ **FIG. 3.5** Internal programmable circuit arrangement for 16 of the i/o pins.

Experiment#1. To read the digital voltage state inputted to P0, (0 V<V_{IN}<+5V)
Wire the pot to give a variable voltage at P0 and use the program below to display a 0 or 1 on the PC
screen. Use a multimeter to measure, V_{IN}.
(Not necessary to control output state switch so don't care about "out0". You can put 0 or 1)

```
          DIR0 = 0                'set pin0, P0, to input
again:    DEBUG DEC IN0,CR        'show P0 logic level on pc screen
          PAUSE 100               'slow down the measurement to 10 Hz
          GOTO again             'keep measuring
```

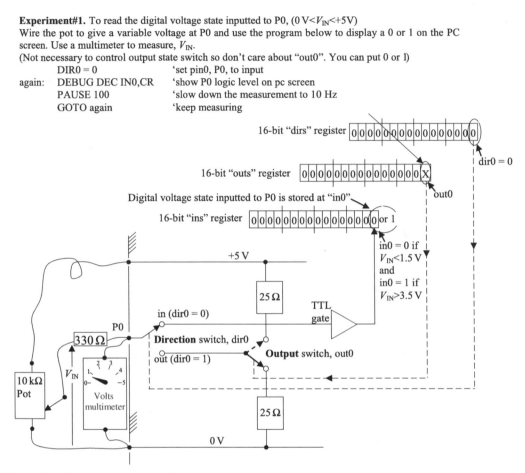

■ **FIG. 3.6** Reading a low and high input state on P0.

itself. This regulated power supply is very useful for powering sensors on
your robot so long as you do not exceed the maximum specified current.
Also connect a 330 Ω resistor at the input of P0 as current protection just
in case DIR0 has been set to 1, that is, output. This concept will be explained
later in this chapter. You will also need a multimeter to measure volts. You
can play with the potentiometer to vary the voltage, V_{IN} applied to P0 from 0
to +5 V. Now write and download the program in Fig. 3.6. Notice that the
program first line sets the direction of P0 to input, DIR0=0. The next
line reads the input logic status of IN0 with the DEBUG instruction. The
DEBUG DEC IN0 instruction shows the value of IN0 as a decimal number
on the pc screen. You can also write DEBUG BIN IN0 and show the result as
a binary number. However, the result will be the same because 0 and 1 are

the same in binary or decimal numbers. By the way, the CR instruction means "carriage return" and if you don't use it then the results will spread across the screen which is a mess rather than down the screen which is readable. You will see that IN0 will read 0 with V_{IN} less than 1.5 V and will change to 1 as the voltage increases to approximately 3.5 V. You should do experiments to understand and measure the hysteresis of change of state meaning that what is the value of V_{IN} for IN0 to change from 0 to 1 when the voltage is increasing and, conversely, what is the value of V_{IN} for IN0 to change from 1 to 0? If the V_{IN} values are different then there is hysteresis and if so why is there this effect and is it useful? You can repeat this experiment on another pin, for example, P11 but remember your program should be changed to DIR11 = 0, DEBUG DEC IN11, CR.

3.4 EXPERIMENT #2. SETTING A LOW ON AN OUTPUT PIN

The following experiment, Fig. 3.7, uses code to set a logic low on pin 0. A quicker way of doing this is LOW 0 but the code shown in the experiment is more fundamental.

Experiment#2. To set voltage state outputted from P0 to, '0', and to observe its value. Use a multimeter, figure 3.6 to measure, V_{OUT}, after running the program below which will set a '0', (V_{OUT} = 0V), at P0

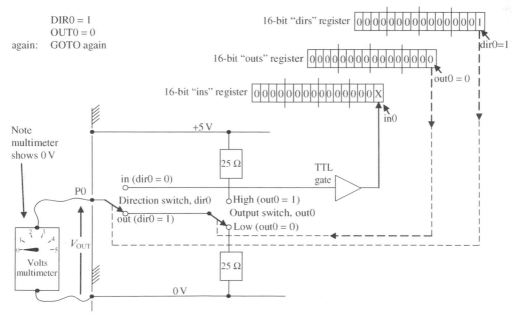

■ **FIG. 3.7** Setting a low output state on pin 0.

Experiment#3. To set voltage state outputted from P0 to, '1', and observe its value of +5V with a voltmeter, figure 3.6.

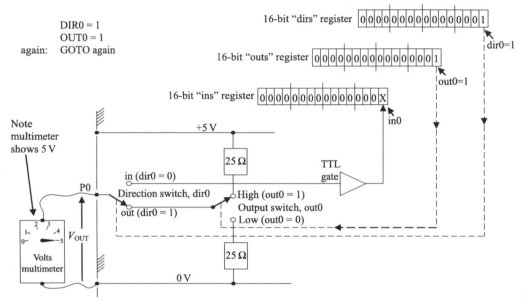

■ **FIG. 3.8** Setting a high output state on pin 0.

3.5 **EXPERIMENT #3. SETTING A HIGH ON AN OUTPUT PIN**

The following experiment, Fig. 3.8, uses code to set a logic high on pin 0. A quicker way of doing this is HIGH 0 but, once again, the code shown in the experiment is more fundamental.

3.6 **SHORT CIRCUIT PROTECTION OF I/O PINS**

It is a really fun and interesting thing to connect Basic Stamps together via their i/o pins and get them to communicate with each other via PULSOUT/ PULSIN and SEROUT/SERIN instructions. The author has designed and built many robots that do this and you can create multicore parallel processing microcomputer architecture systems with their own operating systems, for example, one particular robot was an eight legged spider robot with three degree-of-freedom legs that was controlled by nine interconnected BS2SX microcomputers. Now, if you interconnect Stamps then you should always put a protection resistor between the interconnected pins. A 330 Ω resistor is suitable and the reason for it is shown in Fig. 3.9. Supposing that BS2SX#1 P0 wants to send information to BS2SX#2. P0, then BS2SX#1 P0 is the transmitter and is set to output so it can place zeros and ones on its pin 0.

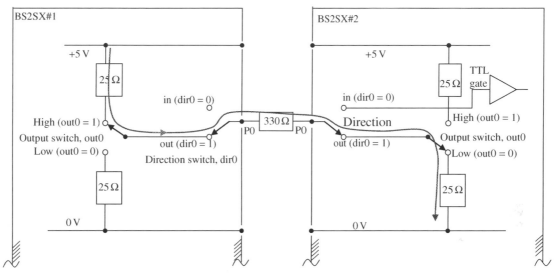

■ **FIG. 3.9** Input/output port protection for the BS2SX microcomputer.

On the other hand, BS2SX#2 is the receiver and must be able to read ones and zeros so it is set to input. Now the output impedance of BS2SX#1 P0 is $25\,\Omega$ and the input impedance of BS2SX P0 is in Megohms so there will be tiny microamps of current flow between pins, which is not a problem. However, it is easy to make a mistake and the BS2SX#2 P0 is set to low impedance output rather than high impedance input. In this case high current can flow and blow up the i/o pins due to the short circuit between the pins. However, this doesn't happen unless the transmitter is set high and the receiver is set low or conversely, the transmitter is set low and the receiver is set high. There are two more combinations which are both set high and both set low and these two will not draw damaging current. However, Fig. 3.9 shows how a dangerous situation can be prevented with a $330\,\Omega$ resistor. The resistor limits the current to $(5\,V - 0\,V)/330\,\Omega = 15\,mA$. The maximum current permissible is $20\,mA$ so no damage will be done.

The next question arises is how will the resistor affect a transmitted high on BS2SX#1 P0. Let's say the input impedance of BS2SX#2 P0 is $1\,M\Omega$ so the voltage transmitted to the receiver is $5\,V \times 1\,M\Omega/(1\,M\Omega + 330\,\Omega + 25\,\Omega) = 4.998\,V$ will be transmitted to the receiver that is well above the $3.5\,V$ threshold for a logic high. So the conclusion is that the protection resistor does its job of protecting and at the same time does not interfere with logic levels and should always be used as a standard procedure when connecting between Stamps. It is also a good idea to do if you are unsure of the impedance of any device that you connect to other than a Stamp.

3.7 **THE RCTIME INSTRUCTION**

The RCtime instruction is a classic analog-to-digital technique for measuring the value of a resistor or a capacitor. It is an instruction that converts an analog quantity into a digital number. The instruction measures the time it takes for a capacitor, C to charge through a resistance, R. The charge time of an RC circuit is proportional to its time constant, $\tau = R \times C$ seconds. The RCtime instruction does not measure the value of resistance in ohms or the value of capacitance in Farads. What it does is to give a digital number that is proportional to these values. To obtain ohms and Farads you must do calibration conversion. If the resistance varies proportionally to, let's say, position, temperature, or light intensity and if the capacitance is a fixed value, then the charge time will represent the value of position, temperature, or light, respectively. Furthermore, if the reflected light intensity is proportional to distance, as is the property of the QTI sensor (Hitting Robot, Chapter 6), then the RCtime instruction can measure range between objects using a noncontact range finding method. If a linear rotary potentiometer is wired as a variable resistor then the rotation angle, θ of the variable resistance wiper will vary with the resistance, R where R is proportional to angle, θ. Thus the RCtime instruction can measure angle and if the variable resistance is a straight-line device rather than rotary then it can measure rectilinear distance. Alternatively, the resistance, R can be a fixed value and the RCtime instruction measures the value of a variable capacitance, C. An extremely sensitive displacement sensor, albeit nonlinear, can be built using C as the variable quantity based on the relationship that capacitance, C is inversely proportional to distance, d between two plates. Another technique is to use the RCtime instruction for an LRtime measurement where L is a variable inductance. The author well remembers developing such an inductive displacement sensor utilizing the LRtime technique with a team in the late 1970s that developed, in the United Kingdom, the world's first automotive microcomputer controlled electronic diesel fuel injection system called the Lucas Epic system.

Ok, let's get down to business. Fig. 3.8 shows the basic structure inside the Stamp of the RCtime instruction. On the left-hand side of the diagram a fixed value capacitor, C is connected in series with a potentiometer wired as a variable resistor, R. The capacitor is connected to +5 V and R is connected to 0 V. A 330 Ω protection resistor is connected between the junction of C and R and fed to P7, the pin that will be used to execute the RCtime instruction.

We are using a BS2SX microcomputer which has a clock speed of 1.25 MHz, that is, a clock period of 0.8 μs. Suppose we invoke the RCtime instruction with the following lines of code:

a	VAR	WORD	'16-bit variable for measuring value of resistance
again:	HIGH 7		'discharge the capacitor by shorting across its terminals
	PAUSE 1		'wait 1ms to allow full discharge
	RCTIME 7,1,a		'measure the value of R and put value in variable a
	DEBUG DEC a,CR		' show the value on the screed as a decimal number
	GOTO again		' go and measure again

This is what happens to the electrical and the digital circuitry inside the Stamp. You need to refer to Figs. 3.10 and 3.11. Fig. 3.10 shows the overview of RCtime and Fig. 3.11 shows actual waveforms measured at V_{in}, V_{gate}, and V_{count} in Fig. 3.10. First of all P0 starts with its direction set to input thus there is a high input impedance at the CR junction. This means that the capacitor, C (0.1 µF capacitance), is fully charged with one side at

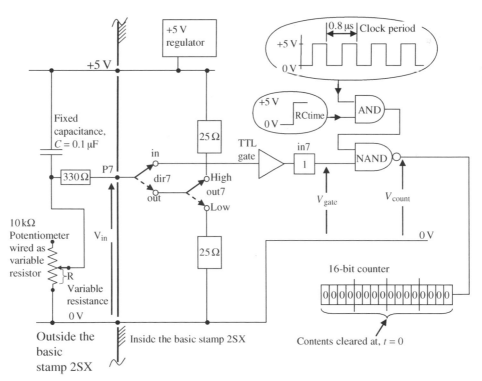

■ **FIG. 3.10** Using the RCtime instruction and the BS2SX internal clock to measure resistance, *R*.

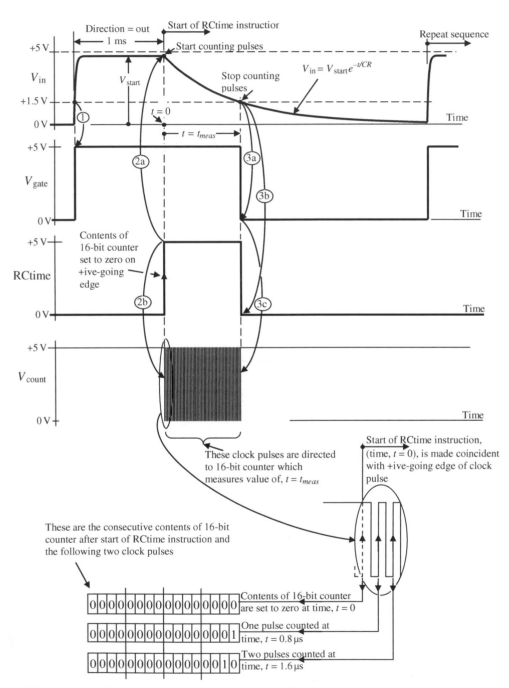

■ **FIG. 3.11** Voltage and clock waveforms and counter values during activation of the RCtime instruction.

+5 V and the other side at 0 V. Next, the HIGH 7 instruction does two things, first, it sets DIR0 direction to OUT (DIR0 = 1) and second, it connects OUT0 to +5 V (OUT0 = 1). What this does is to short circuit capacitor C through the protection resistor which limits the maximum current to 5 V/$(330\,\Omega + 25\,\Omega) = 14$ mA which is well below the 20 mA maximum limit. The time constant of this discharge is $0.1\,\mu F \times (330\,\Omega + 25\,\Omega) \approx 36\,\mu s$. The next instruction is to pause for 1 ms to allow the discharge to fully complete in 1 ms/$36\,\mu s \approx 28$ time constants which is more than enough to fully discharge capacitor, C. You can see the discharge waveform as V_{in} in Fig. 3.11 where the final discharge value measured is $V_{start} = 5\,V \times R/(R + 25\,\Omega)$. For low values of R there may be some nonlinearity so we will assume that R is few hundred ohms. Now, after the 1 ms wait, the RCtime instruction starts and what it does is to initiate charge of the capacitor by removing the short circuit across the capacitor. This is done by setting the direction of DIR0 back to input which allows positive current to flow out of the capacitor plate next to the resistor, R, to flow through R, thence to flow through the power supply thus dropping the voltage, V_{in} back down to 0 V thus charging the capacitor to 5 V across its plates. The time constant for this charge is $R \times C$ and can be seen as V_{in} in Fig. 3.11 identified as the waveform from $t = 0$. Note that the voltage drops as the capacitor charges because it is on the top half of the RC circuit. Students are confused because they expect a waveform to increase as a capacitor charges. The RCtime code is repeated and the waveform repeats.

Now let's go back and see what the Stamp does with the charge waveform. A 16-bit counter is reserved for measuring the digital number that represents the R value. This counter is cleared to zero before the RCtime instruction starts. This counter is fed with pulses that it counts. These pulses come from the onboard $0.8\,\mu s$ clock and are gated by the RCtime instruction and the IN7 TTL gate. Pulses are allowed through the counter only when (i) the RCtime instruction is activated, that is, at $t = 0$ **and** (ii) IN7 is high. Furthermore, IN7 is only high when the charging waveform is above the 1.5 V threshold of the IN7 TTL gate.

The RCtime voltage waveform starts low until the RCtime instruction is initiated at time $t = 0$ when the RCtime voltage goes high. This high value clears the 16-bit counter to zero and permits the AND gate to pass clock pulses to the NAND gate, 2b in Fig. 3.11. At that point in time, 2a in Fig. 3.11, the charge waveform, V_{in} is sent to the TTL gate. If the charge waveform is above 1.5 V then the TTL gate is high and clock pulses are sent to the counter which starts to count pulses. The V_{gate} voltage stays high until $V_{in} = 1.5\,V - V_{start}\ exp(-t/CR)$. At the time when this equation becomes equal to the gate voltage drops to low, 3a in Fig. 3.11, and RCtime voltage drops to low, 3b and clock pulses are stopped from going to the counter, 3c. At this point in time, the value of the counter is loaded into variable a, the RCtime instruction is brought to an end and the program moves on to the next instruction, DEBUG DEC a, which displays the value of a on the screen as a decimal number.

Before we wrap up the discussion on RCtime, we need to explain the "1" in the RCtime 7,1,a instruction. The "1" means to gate the clock pulses through the counter when RCtime is high. The RC circuit can be reversed with the capacitor, C put in the lower grounded half of the circuit and the resistor, R connected to the +5 V rail. In this case the V_{gate} voltage will increase as it is charged. Furthermore, the RCtime instruction has to make other changes too, such as discharge the capacitor by setting OUT7 to low. These changes are automatically carried out by changing the instruction to RCtime 7,0,a. In other words, "1" changes to "0."

3.8 **THE BASIC STAMP AND ITS INTEGER NUMBER COMPUTATION**

The Basic Stamp does not work with decimal point numbers; it works with 16-bit integer numbers, that is, numbers that vary from 0 to 65,535. Furthermore the Stamp can only do integer addition, subtraction, multiplication, and division without decimal points. Students find themselves asking well how does a hand calculator or Excel spreadsheet calculates complicated mathematical functions. This is a worthy research project to find out. The wonderful thing about this is that it causes the students understand numbers much more easily than they would by dealing with numbers with decimal points, trigonometric, and other functions that are easily computed. In fact it turns out in computation that decimal point numbers are simply not necessary. You need to decide on your required precision and work with integer numbers to have that precision. For example, if the required positioning precision of each axis of a three-dimensional (3D) printer that you are designing is 10 μm then 1 mm is represented by 100. There is no need to create a decimal point number of 0.01 mm or 0.00001 m. Decimal point numbers in computation are costly in terms of additional computation. If you are working with Excel spreadsheet then that's fine but if you are working with lean and mean real-time mechatronic and robot machines then you need to be computationally efficient and to do that you need to learn how to work with integers and compute mathematical functions.

3.9 SIGNIFICANT FIGURES AND DECIMAL PLACES

It is important to realize at this point that the number of significant figures is proportional to the accuracy of the number itself and to any calculations in which it is involved. In other words, the accuracy of a number is given by the number of significant figures and *not* the number of decimal places. It is a misdirection to think of the number of decimal places as the quantity giving accuracy to a number or a number calculation. Precisely, the accuracy of a number or number calculation is proportional to the number of places before the decimal point as well as the number of places after the decimal point. That number of places is the number of significant figures. If scientific notation is used, for example, $3.16E-4$ then it can be said that the number of decimal places is an indicator of the precision of a number but that is only because there will always be a nonzero number before the decimal point.

Here is one example: Which number is more accurate? 1234 or 1.23?

The answer is 1234 because it has 4 sig. figs which is a more significant figure than 1.23 which has 3 sig figs. Why is this? 1234 has no decimal places and 1.23 has two decimal places. Well it is because when you specify a number as 1234 it means that it is not 1233 and it is not 1235 because you said it is 1234. This means the number could be anywhere in the range from 1233.5 up to 1234.499 recurring; in other words, 1234 ± 0.5 which is an error ratio of

$$(\pm 0.5)/1234 \times 100\% = 0.041\% \text{ error ratio}$$

A similar analysis for the number 1.23 shows that its error ratio is $(\pm 0.005)/1.23 \times 100\% = 0.41\%$ error ratio, which is 10 times greater than 0.041%.

Conclusion is that the number with 2 dec. pl. has error 10 times the error of the number with 0 dec. pl.

In other words 1234 is more accurate than 1.23.

By the way 1234 can be specified as 1.234×10^3 which has 3 dec. pl.

Another example: Which number is more accurate? 0.362×10^3 or 54?

Ans: 0.362×10^3 because it has more significant figures.

One need to be careful because the number of significant figures is only a guide to the accuracy of a number. For example, which number is more accurate? 1.01 or 99?

Let's take a look.

Error ratio of 1.01 is $0.005/1.01 = 0.0050$ (expressed to 2 sig fig accuracy)

Error ratio of 99 is $0.5/99 = 0.0051$ (expressed to 2 sig fig accuracy)

So the conclusion is that the two numbers are expressed to almost the same accuracy but one has 3 sig figs and the other has only 2 sig figs.

3.10 COMPUTATION OF MATHEMATICAL FUNCTION WITH INTEGERS

We have one more hurdle to jump in preparing number calculations for digital computers. It concerns the fact that all computers have a maximum integer number that they can handle. For example, the biggest integer that Microsoft Excel spreadsheet can handle is a decimal number with 15 sig. fig, that is, the biggest integer you can enter into the spreadsheet is 999,999,999,999,999. By the way we are not talking about the biggest number that Excel spreadsheet can handle; you can experiment to find out yourself that number; the author made it $999,999,999,999,999 \times 10^{293}$ which is 1 less than 10^{308}; we are talking about the biggest integer that is representative of the number of significant figures which in turn is representative of the computation accuracy of the Excel spreadsheet.

The Basic Stamp is much more limited than Excel spreadsheet. As stated before, it can only work with integers (as all personal digital computers) and the biggest integer it can handle is a 16-bit binary number which equates to the decimal number 65,535. In the course of computation, this number must not and cannot be exceeded. Now it is not the time to throw up your arms in dismay. Here is a chance for character building. After programming the Basic Stamp you will be able to more clearly understand numbers and mathematical functions and your problem-solving skills will be enhanced over and above your classmates. Furthermore, it is good for your character and improves your research capabilities, because in the face of adversity, you will turn to using ingenuity to solve problems and not turn to more expensive higher specification tools. You will be leaner, fitter, and smarter. Oh, by the way, just as you thought things were getting bad, they just got worse; I forgot to tell you that the Basic Stamp, apart from being not able to handle decimal point numbers, cannot perform virtually all the functions found on a calculator key panel such as x^n, log, sin, cos, tan. (Actually it can perform with low accuracy the sine and cosine functions but they are not useful in many applications.) It turns out that complex mathematical functions can be computed to high degree of accuracy with only 16-bit, integer-only, adding, subtracting, multiplying, and dividing operations using, for example, truncated Taylor's series.

3.11 **BASIC STAMP EQUATION SETUP PROCEDURE**

Some rules, procedures, and tips:

1. The Basic Stamp carries out calculations from left to right so: $3 \times 2 + 6$ will compute 12 whereas $3 + 2 \times 6$ will compute 30. Multiplication is not given precedence as in handheld calculators.
2. If you divide integers you do not get fractions or decimal point numbers. So, $3/2 = 1$, the 0.5 is lost. Other examples: $32/2 = 16$; $33/2 = 16$; $36/3 = 12$; $35/3 = 11$; $56/57 = 0$; $2/3 = 0$; $57/56 = 1$; and $95/10 = 9$.
3. Be careful of overrunning (=exceeding) 65,535; for example, the calculation, $95 \times 40 \times 40/10$ is calculated in steps from the furthest left number as first $95 \times 40 = 3800$ which is less than 65,535 so that's ok, then, $3800 \times 40 = 152,000$ which is greater than 65,535 so that's not ok since we have overrun. The solution is to change the order of multiplication and division like this: $95 \times 40/10 \times 40$ which gives the sequence of number calculations as $95 \times 40 = 3800$ then $3800/10 = 380$ and then $380 \times 40 = 15,200$.
4. Do multiplication first and division last. For example, $95 \times 40/10 = 380$ but not $95/10 \times 40 = 360$ which is wrong. This is because the sequence of computation is $95/10 = 9$ then $9 \times 40 = 360$.
5. For improved accuracy multiply the biggest numbers first. For example, $95 \times 93 \times 41/10$. The correct answer is 36,223.5, but here are the three combinations of doing Basic Stamp computations:

$95 \times 93/10 \times 41 = 883 \times 41 = 36,203$	error $= 36,223.5 - 36,203 = 20.5$
$95 \times 41/10 \times 93 = 389 \times 93 = 36,177$	error $= 36,223.5 - 36,177 = 46.5$
	(why the greatest error?)
$41 \times 93/10 \times 95 = 381 \times 95 = 36,195$	error $= 36223.5 - 36,195 = 28.5$

Hence the least error is produced by multiplying the biggest numbers first, with the proviso that 65,535 is not exceeded, then divide later in stages.

3.12 **PROBLEMS**

1. Use the RCtime instruction to calibrate accurately the value of R as a function of RCtime digital number.
2. Find the mathematical function for predicting the digital number for a CR circuit.
3. How does a hand calculator compute trigonometric functions, for example, $\sin x$.
4. Calculate $\sin(\theta°)$ from 0 to 90 degrees in 1 degree increments to 1% error using the Basic Stamp.

Chapter

4

Theory III: The Stepper Motor and Its Control

LEARNING OUTCOMES

1. Perception of the physical construction of a hybrid permanent magnet stepper motor.
2. Understanding basic electromagnetic production of sequenced displacement and force.
3. Knowledge of stepper motor electrical power drivers and sequencing.
4. Appreciation of speed and torque limitations of an open-loop stepper motor.

4.1 INTRODUCTION TO THE STEPPER MOTOR AND ITS POWER DRIVERS

The stepper motor, Long's Model 57BYGH7630, Fig. 4.1, is a hybrid-type stepper motor that works with a permanent magnet and soft iron rotor and stator. There are two more types of stepper motor: one is called a "variable reluctance" stepper motor that works only with a soft iron rotor and stator and is without any permanent magnet and the other is called a permanent magnet stepper motor that works with a permanent magnet rotor and a soft iron stator but does not have a soft iron rotor. However, we will only be concerned with the hybrid type here. Stepper motors fall into the family of electric motors called brushless direct current, DC, electric motors, also known as BLDC electric motors. Such motors rely on a DC power supply that supplies current to a sequenced switched current power driver such as the Keyes L298 driver, Fig. 4.2, or the more sophisticated DM542 Fully Digital Stepper Drive, Fig. 4.3, from Leadshine Technology Co. Ltd. Interestingly, the stepper motor can also be considered as belonging to the family of electric motors called alternating current, AC synchronous motors because the stepper motor drivers convert DC electric current into AC current. This AC current, in its basic form, consist of rectangular waveforms that represent the fundamental frequency of the AC waveform. The sophisticated driver,

Creating Precision Robots. https://doi.org/10.1016/B978-0-12-815758-9.00004-3

■ **FIG. 4.1** The Longs model no. 57BYGH7630 hybrid permanent magnet Stepper Motor, shown above. It is a 4-phase electric motor, with rated current of 3.0 A/phase, and step resolution of 1.8 degrees/step, i.e., 20 steps/rev. It is an ingenious, low-cost, precision-made piece of engineering that sells for US$25.

■ **FIG. 4.2** The Keyes L298 stepper motor bipolar power driver shown upper right. This basic driver relies on a 4-phase switching sequence from a microcomputer.

the DM542, can drive a stepper motor with a waveform called a "microstepping" waveform and such a waveform closely matches an AC synchronous waveform and as such the stepper motor can be considered an AC synchronous motor. Fig. 4.4 shows how the stepper motor is driven as a system which consists of: (i) a microcomputer providing control signals,

■ **FIG. 4.3** The Leadshine DM542, shown lower right, is a bipolar microstepping power driver. It is a sophisticated controller that can control the coil current and microstep from 400 step/rev up to 25,600 steps/rev to give very smooth operation. Only two control signals are required; the motor rotation direction, i.e., CW or CCW and the step signal which moves the motor shaft through one-step angle increment.

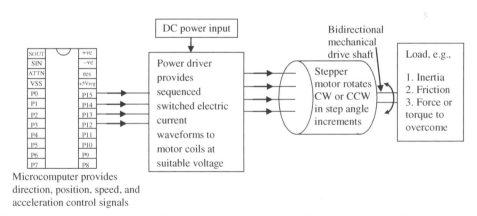

■ **FIG. 4.4** Schematic diagram of stepper motor system that includes a microcomputer, a power driver, the stepper motor, and its load.

(ii) the stepper motor power driver which provides power current signals, and (iii) the stepper motor coils.

In this book we are only concerned with the stepper motor shown in Fig. 4.1, so, from now on we will refer to the Long's 57BYGH7630 hybrid-type stepper motor as the "stepper motor." The stepper motor will now be described in some detail. We will not repeat what is commonly found by a stepper

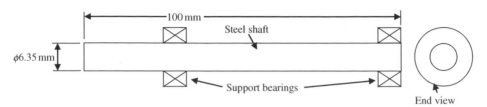

■ **FIG. 4.5** Steel axle plus bearings.

motor internet search. Instead, we will describe the stepper motor more precisely by showing its electromagnetic, electrical, and mechanical design, together with its construction scheme. We start with its construction scheme by discussing the components that makeup the stepper motor.

4.2 MAIN COMPONENTS MAKING UP THE STEPPER MOTOR

Component 1. The Axle

Component 1, Fig. 4.5, is a steel axle on which the permanent magnet system of the stepper motor will be mounted. The axle is supported by a pair of deep groove ball bearing races. The axle is precision ground to a diameter of ¼ in which is equal to 6.35 mm.

Component 2. The Rotor Pole Piece

A rotor, as its name implies, is a component that rotates and produces mechanical output power.

Component 2, Fig. 4.6, is a 50-tooth rotor soft iron pole piece made up from 20 pieces of 0.5 mm thickness laminated, magnetically permeable,

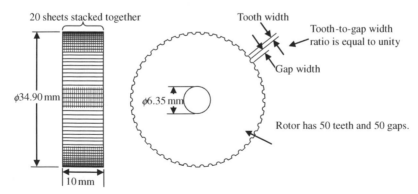

■ **FIG. 4.6** 50-Tooth rotor soft iron pole piece made up from 20 pieces of 0.5 mm thickness laminated magnetically permeable, silicon iron sheets that are riveted tightly together (rivets not shown). Each sheet has thin film of electrically insulated paint on its mating surfaces that reduce eddy current iron losses.

Silicon-iron, SiFe, sheets that are riveted tightly together (rivets not shown). "Soft iron" does not mean that the iron is soft in the sense that a cushion is soft; soft iron means that the magnetic permeability is high which means that the iron will easily conduct a magnetic field. Each of the sheets is painted with a thin coating of electrically insulated paint on either side of its two flat surfaces in order to minimize the eddy current losses. The tooth-to-gap width ratio is equal to unity. The rotor laminations are precision made to an outside diameter of 34.90 mm. The rotor fits inside a stator which is precision made to an inside diameter of 35.00 mm. Thus the rotor-stator air gap is a very small value of $(35.00 - 34.90)/2 = 0.05$ mm thus producing a low reluctance air gap path and a high change in inductance from tooth to gap and vice versa. A high change in inductance as the rotor rotates inside the stator produces a high torque output on the rotor steel axle output shaft. However, this also leads to high cogging torque which leads to noise and vibrations. This disadvantage can be reduced by microstepping which is the advantage of the DM542 driver, Fig. 4.3.

Component 3. The Permanent Magnet

Component 3, Fig. 4.7, is a rare earth, Neodymium Boron Iron, NdBFe, permanent magnet. It is a very magnetically powerful magnet that is axially polarized across its very thin thickness of 2 mm. It is 32 mm in diameter and has a hole in its center that allows it to be threaded onto the steel axle. The permanent magnet is used to polarize the soft iron pole pieces thus producing an extended and specially shaped permanent magnet. It would be

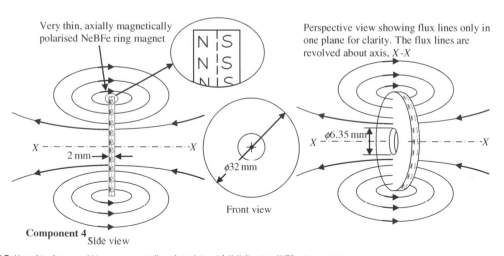

■ **FIG. 4.7** Very thin, 2 mm × ϕ32 mm, magnetically polarized in axial *X-X* direction, NdBFe ring magnet.

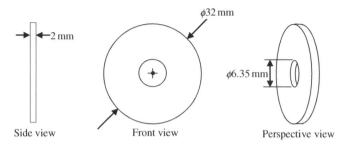

■ **FIG. 4.8** Thin, aluminum disc separator.

expensive, impractical, and unnecessary to shape the permanent magnet to have teeth and gaps for the rotor so the solution is to use easily shaped soft iron for the pole pieces and to match the pole pieces with an easily manufactured permanent magnet such as a cylindrical or rectangular brick shape.

Component 4. The Disc Spacer

Component 4, Fig. 4.8, is a magnetically nonpermeable aluminum disc spacer of the same dimensions as the permanent magnet. We will see shortly that the stepper motor is a twin ganged motor meaning that it has two separate stepper motors mounted on the same axle as if it were two smaller stepper motors connected together producing twice the torque. The nonpermeable aluminum spacer serves the purpose of separating the two stepper motor systems. It is nonpermeable so as not to conduct or upset the magnetic field in the rotor. It has the disadvantage of causing eddy currents due to its high electrical conductance but the magnetic leakage is assumed to be low and thus eddy current losses would be low.

Component 5. The Stator

A stator, as its name implies, is a component that stays static. It forms the body of the stepper motor and a torque is produced on the rotor by a magnetic force interacting between the rotor and the stator.

Component 5, Fig. 4.9, shows the stator that is assembled from 110 pieces of soft iron, SiFe, 0.5 mm thickness sheets that form a stator that is 55 mm long. The sheets are made from the same material as the rotor and likewise are coated with a thin electrical insulated paint that reduces eddy current losses significantly. The end view of stator, Fig. 4.9, is the basic shape that is stamped from the SiFe sheets. Of these shapes 110 are stamped out and laid on top of each other and then riveted tightly together, rivets not shown. The stator, painted black, can be seen in Fig. 4.1. Note the laminations can be

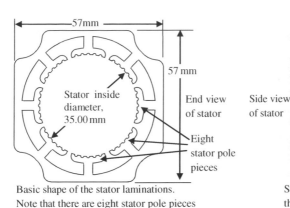

Basic shape of the stator laminations.
Note that there are eight stator pole pieces

Stator assembled from 110 laminated 0.5 mm
thickness sheets of silicon iron. These sheets
are riveted together (rivets not shown).

■ **FIG. 4.9** Diagram of the stator assembly.

seen in the photograph. The inside diameter of the stator is precisely
stamped to 35.00 mm so that when the 34.90 mm outside diameter rotor
is inserted into the stator, the air gap is 0.05 mm. Note that there are eight
stator pole pieces and each pole piece has six teeth with identically shaped
and sized teeth and gaps as the rotor. Thus the stator has a total of
$8 \times 6 = 48$ teeth that interact with 50 teeth on the rotor. The reason will
be explained shortly. Fig. 4.10 shows a view looking down into the center
of the stator. Here, the 55 mm long prismatic, constant cross section stator
shape can be seen, which is created by the many stacked SiFe laminations.

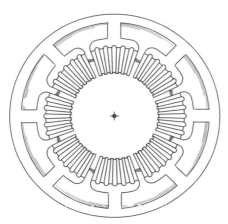

■ **FIG. 4.10** Looking down inside the stator of the stepper motor in the axial direction with the wire coils of
and the stator removed. The stator is shown with a circular outside profile as compared to the square
profile of Fig. 7.9. A circular profile simplifies the stator geometry in order make it easier to clarify flux paths.

We now conclude the descriptions of the fundamental electromagnetic components of the stepper motor. The next step is to assemble the components.

4.3 **ASSEMBLY OF THE STEPPER MOTOR COMPONENTS**

Fig. 4.11 shows the rotor assembly that is composed of axle, two magnets, two pairs of pole pieces, and one separating aluminum spacer. The stepper motor is a dual ganged stepper motor that shares one axle. The four pole pieces are force fitted in an axial direction, with the magnets and separator sandwiched between the pole pieces, onto the axle. During the force fitting process, the pole pieces are precisely oriented in a radial direction such that the pole pieces are oriented tooth, gap, tooth, gap or gap, tooth, gap, tooth when viewed in the axial direction as shown in Fig. 4.12. This arrangement of the teeth and gaps will be explained shortly. Note that the rotor has one pair of rotor pole pieces per rotor gang whereas the stator has four pairs of pole pieces. The reason for this will be explained later.

Fig. 4.13 shows a cross section of the rotor and stator together with a corresponding end view of the two parts. The cross section shows the flux path that circulates from the North poles of the rotor permanent magnets through the rotor North pole pieces then into the soft iron stator pole pieces followed by the outer stator section then entering the South pole of the rotor South pole pieces and returning to the South poles of the permanent magnets. Hence it can be seen that flux return paths are largely axial in their direction.

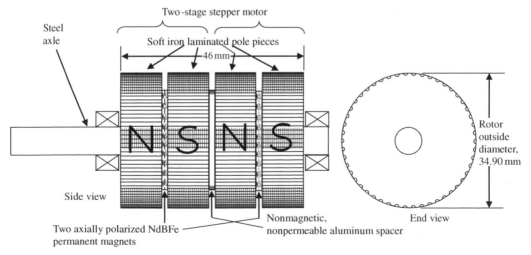

■ **FIG. 4.11** Stepper motor rotor assembly, side view, and end view.

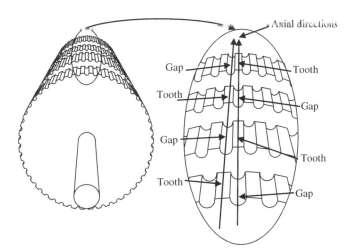

■ **FIG. 4.12** The rotor viewed axially. In the axial direction the rotor has its teeth set at 180 degrees phase difference, i.e., tooth, gap, tooth, gap, or, gap, tooth, gap, tooth.

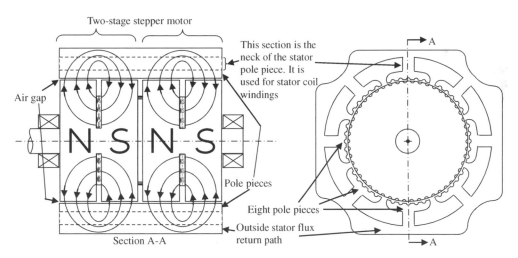

■ **FIG. 4.13** Cross-section through rotor, (left diagram), and end view of stator, (right diagram), showing flux path in cross-section. Note that the stepper motor has two stages.

So far we have not mentioned the gray/silver colored aluminum end caps of the stepper motor, see Fig. 4.1, placed at each end of the black colored stator. These end caps house the two bearings that support the steel axle and rotor assembly but do not play a part in electromagnetic torque production. The end caps, however, are critically important for the precise and stable location of the rotor with respect to the stator due to the very small air gap of 0.05 mm.

We now conclude the description of the stator and the rotor assembly. The next step is to introduce the magnetomotive force, mmf, production brought about by coil windings around the necks of each of the eight stator pole pieces, Fig. 4.13.

4.4 COIL WINDING AND ELECTROMAGNETIC POLARITY

Before we describe mmf production and the coil winding methodology, some basic concepts will be explained. The 8 pole pieces of the stator are wound with 16 elemental coils, 2 elemental coils per pole piece. Each elemental coil, Fig. 4.14, has 35 turns, a resistance of $0.25\,\Omega$ at room temperature, and coil inductance of $0.33\,\text{mH}$. A pair of these elemental coils is wound onto each stator pole piece such that it forms a bifilar coil winding which means that two wires are wound at the same time giving 35 turns per coil.

Considering just one coil, Fig. 4.14, we mark the coil with an arrow. The direction of the arrow indicates the current direction to produce a North pole in the direction of arrow. When current flows in a given direction, the North pole is identified with a black dot, which can be seen in Fig. 4.15. If the electrical current is reversed, Fig. 4.16, then the black dot, indicating the North pole is at the other end of the coil, that is, at the tail end of the arrow.

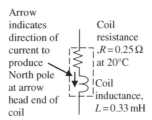

Arrow indicates direction of current to produce North pole at arrow head end of coil

Coil resistance $,R = 0.25\,\Omega$ at 20°C

Coil inductance, $L = 0.33\,\text{mH}$

■ **FIG. 4.14** Elemental coil. There are 16 of these coils that make up the stepper motor.

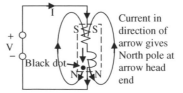

I

$+$
V
$-$

Black dot

Current in direction of arrow gives North pole at arrow head end

■ **FIG. 4.15** Elemental coil with current in +ve reference direction as marked by the arrow.
This direction will give North pole at arrow head end of coil. A black dot is used to identify the North pole.

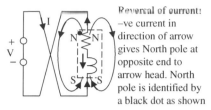

Reversal of current:
−ve current in direction of arrow gives North pole at opposite end to arrow head. North pole is identified by a black dot as shown

■ **FIG. 4.16** Elemental coil with current in −ve reference direction. This direction will give South pole at arrow head end of coil. Once again, a black dot is used to identify the North pole.

4.5 **STATOR AND ROTOR POLE PIECE ATTRACTION**

The rotor is motivated to rotate by attractive magnetic forces acting in a *tangential* direction at the periphery of the rotor. These tangential forces produce a torque on the rotor that is transferred to the axle. The principle is to switch the stator coil currents to attract the rotor to a new angular position. At this new position the torque on the rotor is zero. However, if it is deflected a little either side of that angle with an applied clockwise (CW) or counterclockwise (CCW) rotor torque then the rotor will experience a "restoring torque" that will try to pull the rotor back to its zero torque angular position. The applied torque must not exceed a certain value otherwise the rotor will "lose lock" and slip to an unknown position. In order to move the rotor to a new angle, the stator coils are switched to a new status and the rotor will shift to a new angular position. Figs. 4.17 and 4.18 show how the stator coils and stator pole pieces hold the rotor in a given angular position. The coils are not shown for the sake of clarity. Instead the mmf direction is shown.

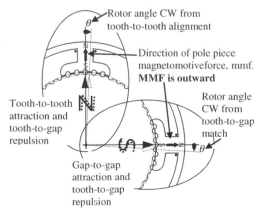

■ **FIG. 4.17** Attractive torque by tooth-to-tooth and gap-to-gap alignment with mmf *outward*.

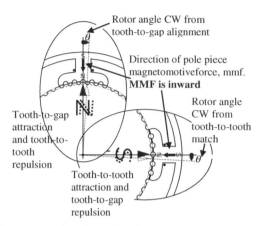

Rotor angle CW from
tooth-to-gap alignment

Direction of pole piece
magnetomotiveforce, mmf.
MMF is inward

Rotor angle
CW from
tooth-to-tooth
match

Tooth-to-gap
attraction
and tooth-to-
tooth
repulsion

Tooth-to-tooth
attraction and
tooth-to-gap
repulsion

■ **FIG. 4.18** Attractive torque by tooth-to-tooth and gap-to-gap alignment with mmf *inward*.

A rotor North pole and a rotor South pole have two opportunities each to attract themselves to a given angular position with respect to the stator. Thus we have four opportunities for pole piece attraction, which are as follows:

1. Rotor North pole tooth is attracted to stator South pole tooth, Fig. 4.17 upper diagram.
2. Rotor South pole tooth is attracted to stator South pole gap, Fig. 4.17 lower diagram.
3. Rotor North pole tooth is attracted to stator North pole gap, Fig. 4.18 upper diagram.
4. Rotor South pole tooth is attracted to stator North pole tooth, Fig. 4.18, lower diagram.

The torque that acts on the rotor for a given stator coil current status is shown in Fig. 4.19. As the rotor deviates from its quiescent zero torque position it experiences a restoring torque. For example, if the rotor is disturbed in the CW direction then there will be a −ve CW torque, that is, a CCW torque, acting on the rotor to try to restore it to its quiescent position. Likewise if the rotor is disturbed in the CCW direction then there will be a CW restoring torque. The further the rotor deviates from the quiescent position the greater the restoring torque. The effect is to replicate a mechanical spring. However, there comes a point when the restoring torque reaches a maximum at approximately ¼ tooth pitch on either side of the quiescent position and the slope of rotor torque against rotor angle starts to become negative. This means that the rotor position becomes unstable meaning that the rotor will rotate uncontrollably. This phenomenon is known as "losing lock" and will be analyzed later in more detail with "pull-in" and "pull-out" tests.

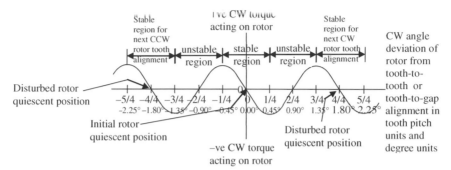

■ **FIG. 4.19** CW torque acting on rotor versus angle deviation from tooth-to-tooth alignment and tooth-to-gap alignment.

Note from Fig. 4.19 that the next position that is stable for the rotor is at ± 1 tooth pitch away from the quiescent position. This means that when the stepper motor is switched on, the rotor will take up an unknown position which is the major disadvantage of the stepper motor. It is because a stepper is an open-loop device meaning that the rotor shaft angle is not measured. The problem is overcome by driving the rotor to a known position, measured by a sensor and then the stepper motor is driven in a specified profile by counting the number of steps CW or CCW. Care must be taken so that the stepper motor is neither accelerated too quickly nor stepped at too high stepping rate. These concerns are examined later by undertaking experiments on the stepper motor to determine its "pull-in" and "pull-out" stepping rates in order to drive the stepper motor effectively.

In order to achieve the four opportunities for pole piece attraction as listed above, it is necessary to reverse the mmf in the stator pole pieces. Fig. 4.17 shows the mmf vectored outward and Fig. 4.18 shows the mmf vectored inward. This means that we need a bipolar current driver in order to sequence the stator coil currents to produce rotor rotation. The next step in understanding the stepper motor, and before introducing the current sequencing power driver circuit, is to explain the coil winding methodology around the eight salient stator pole pieces. Note that "salient" means "to stick out," "to protrude."

4.6 COIL WINDING METHODOLOGY

Fig. 4.20 shows the coil winding methodology of the stepper motor in a geometrical layout corresponding to the shape of the stator. Fig. 4.21 shows the stator pole piece windings in a physical context. From the figures a pair of elemental coils wound in a bifilar arrangement around each stator salient

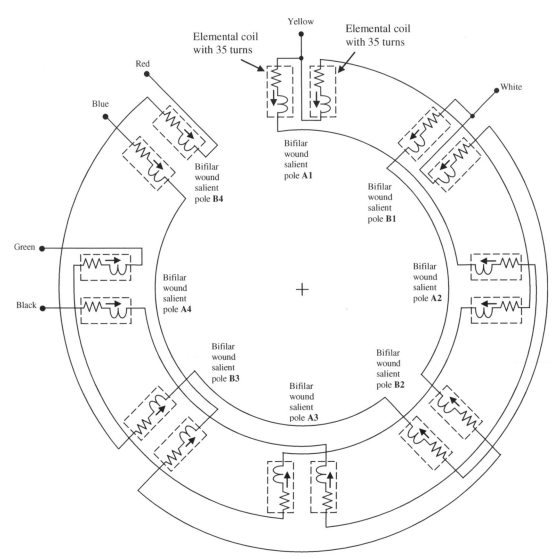

■ **FIG. 4.20** Electrical wiring diagram, of the 8 salient stator poles of the stepper motor. Note that each salient pole has a pair of bifilar wound coil windings making 16 elemental coils in total.

pole can be seen. Since there are 8 stator salient poles, there are 16 elemental coils. Note that every alternate salient pole coil is connected, that is, A1 is connected to A2 is connected to A3 is connected to A4 and likewise, B1 is connected to B2 is connected to B3 is connected to B4. In other words there is no connection between adjacent coils. The connection system is ingenious and is best shown by a circuit diagram in Fig. 4.22.

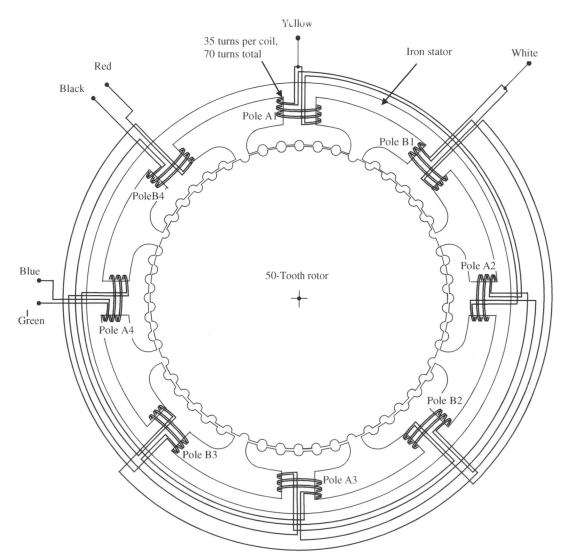

Yellow

35 turns per coil, 70 turns total

Iron stator

White

Red

Black

Pole A1

Pole B1

PoleB4

Pole A2

Blue

50-Tooth rotor

Green

Pole A4

Pole B2

Pole B3

Pole A3

■ **FIG. 4.21** Electrical wiring diagram, shown in a physical context, of the eight salient pole coil windings of the stepper motor. Note that each salient pole has a pair of bifilar wound coil windings making 16 coils in total. The diagram is tricky to follow. A sharp pencil is needed to follow each wire from its starting point to its destination. Be careful to distinguish between wires and the iron stator.

The wiring system allows a unipolar drive power drive circuit or a bipolar drive power drive circuit. The unipolar drive circuit requires the use of the yellow and white connections as shown in Figs. 4.21 and 4.22 but the bipolar drive does not. It turns out that the unipolar drive circuit nowadays is only useful as an academic exercise. The authors devised a unipolar circuit on one

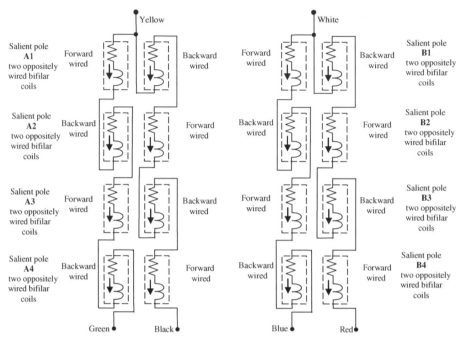

■ **FIG. 4.22** Electrical wiring diagram of the 16 elemental coils of the stator salient pole piece windings. Note the ingenious winding arrangement. There are eight bifilar wound coils. The bifilar wound coils are wired in opposition as "backward wired" and "forward wired." The winding arrangement means that a unipolar as well as a bipolar switching sequence is possible.

occasion to demonstrate to students, and this circuitry is shown later, but the bipolar circuit has really rendered the unipolar circuit obsolete. This is because of the improvements in the performance and lowering of the cost of transistor switching integrated circuit technology, for example, the Keyes driver, Fig. 4.1. The unipolar method of driving is inefficient because only one quarter of the windings is being used at any time to produce torque. Also power loss resistors are required so that the coils can be energized by a high supply voltage to give high performance and this leads to power loss in the power resistors. Thus the unipolar method of driving is inefficient both electrically and electromagnetically. The main advantage of the unipolar drive is that only four transistors are required. Actually, the power loss resistors can be made redundant with the use of a chopper drive circuit. The bipolar circuit method of driving, on the other hand, when used with a current control chopping circuit is highly efficient and has high performance because there are no power loss resistors, the coils are utilized 50%, and the mechanical output power is increased.

Fig. 4.22 shows a circuit diagram of the bifilar windings that are wound onto each salient stator pole piece. Here it can be seen that each bifilar winding is wired in opposition, as "forward" wired and "backward" wired. The reason for this is shown in Fig. 4.23. Figs. 4.23A and B show two types of unipolar connection of a set of stator windings either winding set, A1, A2, A3, A4 or winding set, B1, B2, B3, B4. Similarly, Figs. 4.23A and B show two types of bipolar connection. These figures show that the unipolar connection uses only half of the available set of windings whereas the bipolar connection uses the whole set of windings. Furthermore, the windings, whether connected as unipolar or bipolar, produce alternate directions of magnetomotive force, mmf that will produce appropriate attraction of pole pieces to be shown shortly.

Before we discuss the electromagnetic force interaction between rotor and stator pole pieces, the geometrical design needs to be explained. First of all, the rotor has 50 teeth, equispaced with gaps, in other words, 50 teeth and 50 gaps with a unity tooth-to-gap ratio, Figs. 4.6 and 4.24. The stator has eight equispaced pole pieces, that is, each pole piece is separated by 45 degrees. Each stator pole piece has 6 teeth that identically match the rotor teeth and gaps. Thus stator thus has $8 \times 6 = 48$ teeth on all its pole pieces. The rotor has 50 teeth so the stator has 2 teeth "missing." Since there are

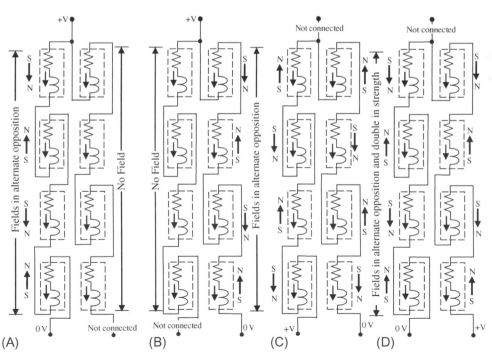

■ **FIG. 4.23** Production of magnetic field using unipolar and bipolar connections of the windings. (A) Unipolar connection #1. (B) Unipolar connection#2. (C) Bipolar connection#1. (D) Bipolar connection#2.

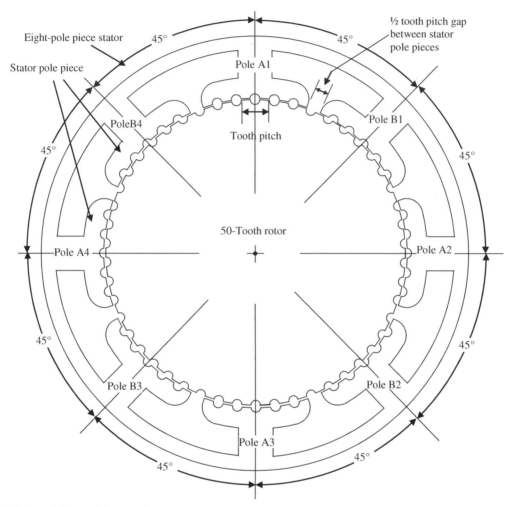

Eight-pole piece stator

Stator pole piece

45°

½ tooth pitch gap between stator pole pieces

Pole A1

PoleB4

Pole B1

Tooth pitch

45°

45°

45°

50-Tooth rotor

Pole A4

Pole A2

45°

45°

Pole B3

Pole B2

Pole A3

45°

45°

■ **FIG. 4.24** Geometrical layout of the rotor and stator.

eight gaps between stator pole pieces, the missing pole piece teeth amounts to ½ tooth pitch between stator pole pieces, Fig. 4.24.

Now refer to Figs. 4.24 and 4.25 top left, and note that the rotor is shown with six of its teeth aligned with the six teeth of the A1 stator pole piece. Due to the stator-rotor geometry, the rotor teeth also align with stator pole piece A3, which is diametrically opposite to A1. At the same time, six rotor *gaps*, not teeth, align with the six teeth of stator pole A2 and A4, in other words there is a 180 degrees phase difference alignment with poles A1 and A3. If, now, the rotor is rotated ¼ tooth pitch CW, that is, by 1.8 degrees, then six rotor teeth will align with teeth of poles B1 and B3 and six rotor gaps will align with the teeth of poles B2 and B4. If the rotor rotates ¼ tooth pitch CCW then rotor tooth gap alignment of B2, B4 is switched with B1, B3.

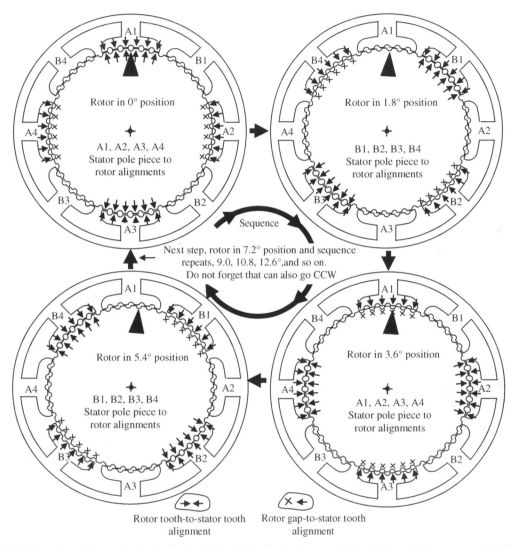

Rotor in 0° position

A1, A2, A3, A4
Stator pole piece to
rotor alignments

Rotor in 1.8° position

B1, B2, B3, B4
Stator pole piece to
rotor alignments

Sequence

Next step, rotor in 7.2° position and sequence
repeats, 9.0, 10.8, 12.6°, and so on.
Do not forget that can also go CCW

Rotor in 5.4° position

B1, B2, B3, B4
Stator pole piece to
rotor alignments

Rotor in 3.6° position

A1, A2, A3, A4
Stator pole piece to
rotor alignments

Rotor tooth-to-stator tooth
alignment

Rotor gap-to-stator tooth
alignment

■ **FIG. 4.25** Rotor rotation step increments showing the rotor tooth-to-stator tooth and rotor gap-to-stator tooth alignments that occur every 1/200 of a revolution, i.e., every 1.8 degrees.

If, now, the rotor is rotated a further 1.8 degrees CW, from rotor position 0 degrees to rotor position 1.8 degrees, Fig. 4.25 top right, then there will rotor-tooth-to-stator-tooth alignment with B1 and B3 and rotor-gap-to-stator-tooth with B2 and B4. The third of fourth combinations of tooth and gap alignment occurs at rotor position 3.6 degrees, Fig. 4.25 lower right and the fourth combination at rotor position 5.4 degrees, Fig. 4.25 lower left. These rotor positions are actuated by electromagnetic sequenced forces that will be explained next. The sequencing means that the stepper motor rotor can be driven CW or CCW in 1.8 degrees steps, that is, 200 steps/revolution and thus *the stepper motor is known as a 200 step/rev stepper motor.*

A further ingenious feature of this arrangement of the magnetic forces is that the forces acting on the rotor in the plane of the paper are in static equilibrium thus there are no forces orthogonal to the rotor axis that will excite vibrations and noise or cause bending of the axle and neither will there be forces that close the very small air gap. Even though the sum of the forces, which are large, on the rotor in the plane of the paper, sum up to zero, the *moment* of these forces about the rotation axis do not sum to zero; in fact they add up to assist each other to give the working output rotor torque of 2 N m; this is a further tribute to the designers of the stepper motor. Also, it needs to be said here that the manufacturing engineers deserve a special tribute concerning the low-cost precision manufacture and assembly of the "humble" stepper motor. We now move on to discuss the magnetic flux distribution in the stepper motor.

4.7 FLUX DISTRIBUTION AND HOW IT IS USED TO STEP THE ROTOR IN INCREMENTS

There are two sets of stator coils, Fig. 4.20: coils, A1, A2, A3, A4 and coils, B1, B2, B3, B4. The coils of each set are interconnected, Fig. 4.22. The coils of each set are mounted on the stator at 90 degrees intervals, Fig. 4.21 and each set is mutually 45 degrees apart. The rotor is rotated in steps by the alternate sequencing of bipolar current in each set, that is,

Step 1 (0.0°)	+ve current ON in coils A1, A2, A3, A4;	current OFF in coils B1,B2, B3, B4
Step 2 (1.8°)	current OFF in coils A1, A2, A3, A4;	+ve current ON in coils B1,B2, B3, B4
Step 3 (3.6°)	−ve current ON in coils A1, A2, A3, A4;	current OFF in coils B1,B2, B3, B4
Step 4 (5.4 °)	current OFF in coils A1, A2, A3, A4;	−ve current ON in coils B1,B2, B3, B4

...and repeat the sequence, for example,...

Step 5 (7.2°)	+ve current ON	in coils A1, A2, A3, A4 and current OFF in coils B1,B2, B3, B4

The sequence can be abbreviated as A+, B+, A−, B−, A+, and so on.

The magnetic flux effects of this alternate bipolar sequencing of coil current is shown in Fig. 4.26 where the rotor has been removed for clarity. Take careful note of the shape and direction of the flux streamlines or lines of force. For example, set A coils are energized in steps 1 and 3; streamlines are of the same shape but directions are opposite. Likewise, set B coils are energized in steps 2 and 4. It is important to point out that the flux streamlines, Fig. 4.26, are generated by the stator pole pieces and thus run the whole length of the stepper motor, Fig. 4.10.

Next is to show how the stator flux fields interact with the magnetized rotor pole pieces to increment the rotor through 1.8 degree steps. These stator pole pieces are axially prismatic and produce flux fields that interact with the North and South poles of the rotor pole pieces as shown in Fig. 4.27A–D, which show the rotor being stepped through angles, 0, 1.8, 3.6, and 5.4 degrees, respectively, with stator coils being energized in sequence A+, B+, A−, B−, respectively. The flux fields are three

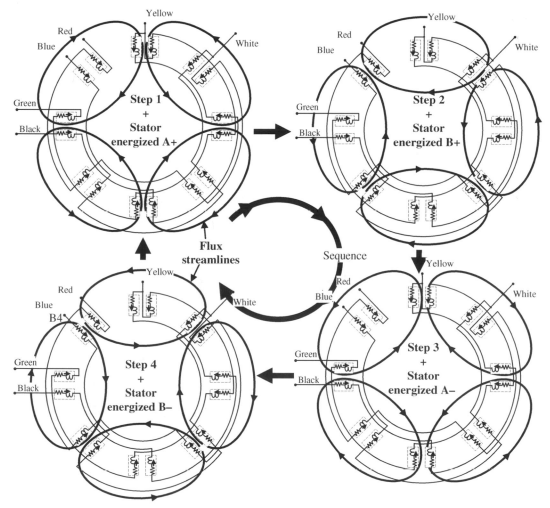

■ **FIG. 4.26** Diagram of the magnetic flux stream lines or "lines of force" with rotor removed when the stator poles are energized in sequence, A+, B+, A−, B−, and repeat. The flux lines are the same whether energized in unipolar or in bipolar mode. The only difference will be that the bipolar mode magneto motive force, mmf, will be twice that of the unipolar mode.

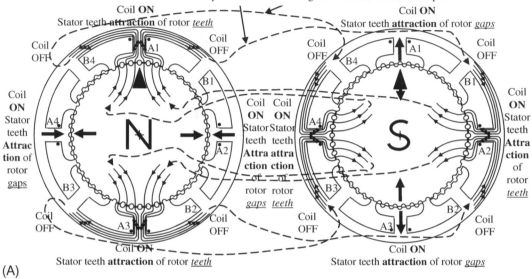

■ **FIG. 4.27A** Stator coil energization. A+ Rotor in 0 degrees position.

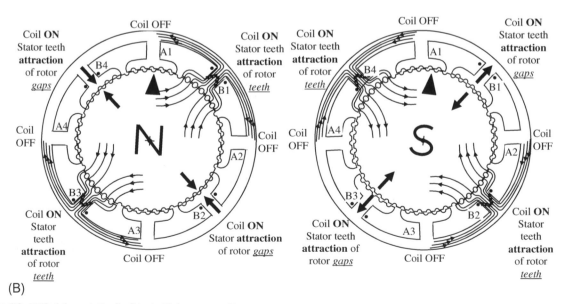

■ **FIG. 4.27B** Coil energization B+ Rotor in 1.8 degrees cw position.

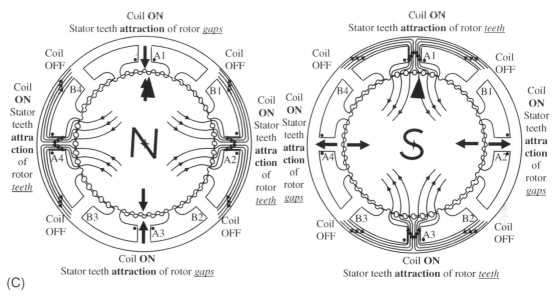

(C)

■ **FIG. 4.27C** Coil energization. A— Rotor in 3.6 degrees position.

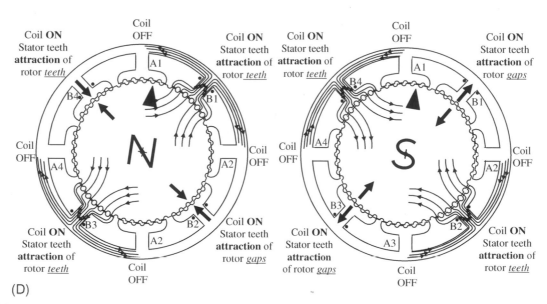

(D)

■ **FIG. 4.27D** Coil energization. B— Rotor in 5.4 degrees position.

dimensional and there is a flux component that "flows" axially along the stator and the rotor, see also Fig. 4.13. Note that there is a lot of information in these diagrams, Fig. 4.27. For example, the two diagrams in Fig. 4.27A, represent one stage of the two-stage stepper motor, (Fig. 4.13 shows two stages), and the North and South pole pieces of the rotor are on the common axle. In reality the two diagrams should be placed on top of each other but they are separated for clarity

Note carefully the status of each coil, whether it is switched on or off and whether the polarity is positive, +ve, or negative, −ve. So, for example, in Fig. 4.27A, left diagram, the A1 stator coil is switched on and has −ve polarity; so also in the right diagram the A1 coil is switched on and has −ve polarity because it is the very same stator pole piece and coil as in the left diagram. Note also the methods of obtaining attraction between the rotor and the stator shown in Fig. 4.25.

We have now completed the description of the electromagnetic torque and stepping motion production. The next stage is to describe the electrical and the electronic switching methodology for stepping the stepper motor through an angle sequence with respect to time.

4.8 STEPPER MOTOR ELECTRICAL POWER DRIVERS

Three driver methods will be described including their circuitry. They are as follows:

(i) The unipolar driver, also known as LR drive.
(ii) The bipolar driver using the L298 chip; incorporated in the "Keyes" circuit board.
(iii) The bipolar driver using the Leadshine Technology, DM542 Digital Driver.

The Unipolar Circuit

Fig. 4.28 shows the circuit diagram of a unipolar driver with a 12 V power supply to it and a Basic Stamp microcomputer that controls the driver. Fig. 4.28 also shows the photograph of the system. It is a straightforward circuit with four International Rectifier enhanced mode mosfet, IRF1414, power transistors, which have low r_{DSon} resistance so do not require a heat sink when conducting 3 A of current which is the rated current per phase of the stepper motor.

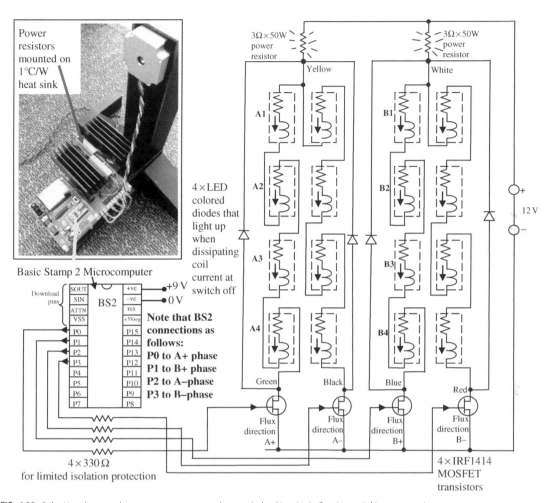

■ **FIG. 4.28** Coil wiring diagram when stepper motor connected as a unipolar drive circuit. Transistor switching sequence is:

Full step	A+,		B+,		A−,		B−
2-Phase-on step		A+B+,		B+A−,		A−B−,		B−A+
Half step	A+,	A+B+,	B+,	B+A−,	A−,	A−B−,	B−,	B−A+

Note, in unipolar mode that "a phase" is the connection between:

  (i) yellow and green, phase Λ+

 (ii) yellow and black, phase A −

(iii) white and blue, phase B+

 (iv) white and red, phase B −

Thus a phase in unipolar mode consists of four elemental coils, see Fig. 4.28. Transistors individually switch on the current in each single phase with a TTL logic high from the Basic Stamp applied to its gate, that is, a 5 V signal. Conversely, a logic low, that is, 0 V, will switch off the transistor. The phase current should not exceed 3 A and the phase resistance is 1 Ω at 20°C so the phase voltage is 3 V. If a 12 V power supply is used, then 9 V should be dissipated in a power resistor. The power resistor thus should have a resistance of 9 V/3 A = 3 Ω and the power rating of $I^2R = (3)^2 \times 3 = 27$ W. We used a power rating of 50 W because that was in our storage rack. Note that the green and black transistors (and the blue and red transistors) are never switched on at the same time so the maximum current taken by any one power resistor never exceeds 3 A. There are three ways or modes that the stepper motor can be driven with a unipolar driver. They are as follows:

(i)   Standard full step mode that gives 1.8 degrees/step and 200 steps/rev.
(ii)   A 2-phase-on step mode that also gives 1.8 degrees/step and 200 steps/rev but with about 40% more torque than full step but the current consumption is twice that of full step mode.
(iii) Half-step mode that is an interweaved mix of full step and 2-phase-on modes that gives 0.9 degrees/step and 400 steps/rev, gives a smoother less noisy motion but consumes 50% more current than full step mode.

4.9 REASON FOR HIGH-VOLTAGE 12 V POWER SUPPLY

The 12 V power supply in Fig. 4.28 is to increase the performance of the stepper motor by increasing the available mechanical output power and by maintaining the motor torque up to an increased stepping speed which relates directly to motor speed, which will now be explained. The stepper motor relies on the switching of current in its coils in order to step CW or CCW. Each coil or phase has an inductance of 1.3 mH and a series resistance of 1 Ω, Figs. 4.14 and 4.28 (note four coils in series per phase). The current in an inductor takes a finite time to rise to its maximum that is limited by the resistance, R. Assuming initial current is zero, the current, I in an LR series circuit is given by

$$I = V/R\left(1 - e^{-t/\tau}\right) \qquad (4.1)$$

where V is the supply voltage, t the time, and τ is the time constant equal to L/R.

Fig. 4.29 shows two circuits and their associated current versus time relationships for switching on current at time $t = 0$. The addition of the 3 Ω series resistor decreases the time constant, τ in Eq. (4.1) four times from

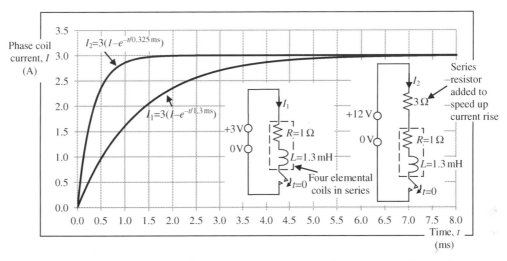

■ **FIG. 4.29** Phase coil current versus time. Note that the addition of the 3 Ω series resistor causes the current rise to be four times faster.

$1.3\,\text{mH}/1\,\Omega = 1.3\,\text{ms}$ to $1.3\,\text{mH}/(1+3)\,\Omega = 0.325\,\text{ms}$, thus the current rise, Fig. 4.29, is four times faster. The drive is sometimes called "L over nR drive" meaning L/nR, where L is phase inductance, R the phase resistance, and n is the number of times the total series resistance of the coil is greater than the phase resistance. In this case $n = 4$. In quantitative terms, it takes approximately 6 ms for the current to rise close to 3 A without the series resistor and approximately 1.5 ms with the series resistor. The down side is the I^2R power loss in the 3 Ω resistor that costs electrical power, dollars, weight, and space due to the power resistors and the heat sink, see Fig. 4.28. Now why is it important to speedup the current rise in the stepper motor coils?

The answer is that coil current directly produces torque, which, when multiplied by the rotor angular velocity, gives the mechanical output power. Fig. 4.30 shows the four coils being switched in unipolar mode sequence A+, B+, A−, B−, A+, *et seq*, with and without a series resistor. Note that the graph is an approximation of the real situation which should take account of the back emf in the coils and the change of inductance as the rotor moves. Nonetheless, it gives an idea of the general principle of improving motor performance.

The coil ON time is 1 ms, so the switching frequency, that is, the stepping frequency, is 1 kHz. It can be seen that less current flows through the coils without an additional series resistor, so the torque will be higher with the resistor. If the stepping frequency is very low, for example, 10 Hz then the increase in performance with the series resistor is negligible. The

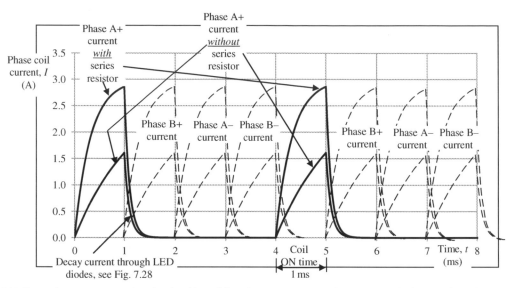

■ **FIG. 4.30** Phase coil current versus time. Note that the addition of the 3 Ω series resistor causes the current rise to be four times faster. This graph is an approximation of the real situation which should take account of the back emf in the coils and the change of inductance as the rotor moves.

purpose of the series resistor is to increase high-speed performance. Fig. 4.30 shows the decay current through the light-emitting diode (LED), see circuit in Fig. 4.28. This decay current is fast due to the high resistance of the LED but nonetheless it will cause a braking effect as the next on phase tries to pull the rotor to its next step thus slowing down the rotor motion.

We now turn out attention to programming stepper motor motion with the Basic Stamp microcomputer.

4.10 BASIC STAMP PROGRAMMING OF UNIPOLAR MODES

1-Phase-on Mode

We will now show how to write Basic Stamp PBasic code to control the transistors in the unipolar circuit shown in Fig. 4.28. You first need to download the Basic Stamp 2 editor on the www.parallax.com website. Here is the code for driving the stepper motor in 1-phase-on unipolar mode, refer to Fig. 4.28. A 1-phase-on mode is a full step mode meaning that the step size is 1.8 degrees giving 200 steps/rev. You will need to learn about the Basic Stamp

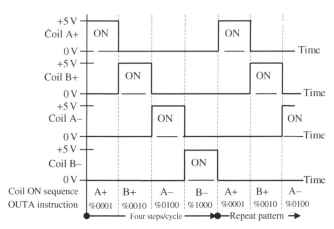

■ **FIG. 4.31** "1-Phase-on" mode voltage waveforms to the transistor gates of the unipolar circuit.

instructions, DIRA, OUTA, and PAUSE, obtainable by clicking the manual at top right in the editor screen. The program produces logic waveforms applied to the unipolar circuit as shown in Fig. 4.31.

Determination of Pull-In Speed

PROGRAM 4.1

```
'{$STAMP BS2}                                  'specify that we are using the Basic Stamp 2 module
'programme to step the stepper motor CW in Full Step mode which gives 200steps/rev
'programme is to determine the "pull-in" rate by decreasing the period until the motor loses lock
'you will know when the motor loses lock because it will be noisy and the shaft will stop turning.
startperiod    VAR    word      'specify a 16 bit integer for the step period
            startperiod=100    ' this variable will set the periodic time of the stepping rate
            DIRA=%1111          'set pins 0,1,2,3 to out direction
            OUTA=%0000          'set pins 0,1,2,3 to low. This turns OFF all transistors
again:      OUTA=%0001              'step 1, set pin 0 to high so as to turn ON phase A+
            PAUSE startperiod      'delay 100millisecs, this will give a 10Hz stepping rate
            OUTA=%0010             'step 2, set pin 1 to high so as to turn ON phase B+
            PAUSE startperiod      'don't forget to put a PAUSE after every phase change
            OUTA=%0100             'step 3, set pin 2 to high so as to turn ON phase A-
            PAUSE startperiod
            OUTA=%1000             'step 4, set pin 3 to high so as to turn ON phase B-
            PAUSE startperiod      'once again don't forget the PAUSE after every phase change
            GOTO again          'repeat sequence for ever
```

Program 4.1 steps the motor at a slow speed of 10 Hz (periodic time 100 ms) in the CW direction. A stepping frequency of 10 Hz means that the rotor turns at $(10 \, steps/s)/(200 \, steps/rev) = 1/20 \, rev/s = 20 \, s/rev$ which is very slow. To go in the CCW direction you reverse the OUTA sequence. To find the pull-in speed, you need to decrease the period value until the rotor loses lock. Then set the period to let us say 10% more than the period that loses lock; this is then termed the "pull-in" speed. The pull-in speed is the maximum speed at which you can safely start the stepper motor moving from rest. Note that the pull-in speed depends on the mass and friction that is connected to the rotor shaft.

Determination of Pull-Out Speed

The following program, Program 4.2, will accelerate the motor from the pull-in speedup to the, yet to be determined, pull-out speed. The pull-out speed is the maximum speed that causes loss of lock. The pull-out speed is greater than the pull-in speed. The pull-out speed is found by careful graduated acceleration of the motor until lock is lost. The pull-out speed occurs due to the decrease in torque as the period decreases, Fig. 4.30, and increase in torque requirement as the speed increases due to increase in viscous and friction torque with speed. Careful graduated acceleration means that the available torque is not wasted too much on the inertial torque.

Pull-Out Speed

PROGRAM 4.2

```
'{$STAMP BS2}
'programme to step the stepper motor CW in Full Step mode from pull-in speed to pull-out speed
'the pull-out speed is to be determined by reducing minperiod until lock is lost
period          VAR     word    'specify a 16 bit integer for the step period
startperiod     CON     100     'CON means a constant
minperiod       CON     10
                DIRA=%1111      'set pins 0,1,2,3 to out direction
                OUTA=%0000      'set pins 0,1,2,3 to low. This turns OFF all transistors
again:          FOR period=startperiod TO minperiod        'ramp up speed
                OUTA=%0001
                PAUSE period
                OUTA=%0010
                PAUSE period
                PAUSE period
                OUTA=%1000
                PAUSE   period
                NEXT
```

```
FOR period=minperiod TO startperiod   'ramp down speed
OUTA=%0001
PAUSE period
OUTA=%0010
PAUSE period
PAUSE period
OUTA=%1000
PAUSE   period
NEXT
GOTO again
```

Program 4.2 acceleration is not constant. In fact, the Program 4.2 acceleration increases with speed when it should decrease with speed because the torque decreases with speed. This is an area that students should investigate and improve.

2-Phase-On Step Mode

A 2-phase-on mode is also a full step mode since the step size is also 1.8 degrees. This mode always has 2-phase coils switched on at any one time, Fig. 4.32, so its current draw is twice that of the 1-phase-on mode. The torque of the 2-phase-on mode is approximately 40% more than the 1-phase-on mode due to the vector sum of torques. Programs 4.1 and 4.2 should be repeated using 2-phase-on step mode to find the pull-in and pull-out speeds. The sequence using the OUTA instruction is as follows:

```
again:   OUTA=%0011     'step 1   A+ and B+ on
         PAUSE period
         OUTA=%0110     'step 2   B+ and A- on
         PAUSE period
         OUTA=%1100     'step 3   A- and B- on
         PAUSE period
         OUTA=%1001     'step 4   B- and A+ on
         PAUSE period
         GOTO again
```

Half-Step Mode

Once again, Programs 4.1 and 4.2 should be repeated using the half-step mode. Half-step mode has a step angle of 0.9 degrees thus producing 400 steps/rev. Half-step mode has the advantage that it gives quieter and smoother operation with less vibration input to the robot structure. This time we introduce an alternative to the PAUSE instruction that has many

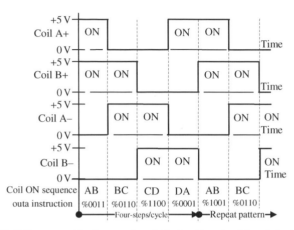

■ **FIG. 4.32** "2-Phase on mode" voltage waveforms to the transistor gates of the unipolar circuit.

times more resolution meaning that the period can be more precise and can be programmed to a lower value. The alternative instruction is PULSOUT which has a resolution of 2 μs rather than 1 ms of the PAUSE instruction. This means that the PULSOUT instruction is 500 times more resolute than the PAUSE instruction. Hence the "period" variable should be multiplied by 500 when compared with the PAUSE instruction. Fig. 4.33 shows the waveforms created by the Basic Stamp. Half-step mode is an alternate interleaved mix of 1-phase-on mode and 2-phase-on mode so the current alternates between 3 and 6 A.

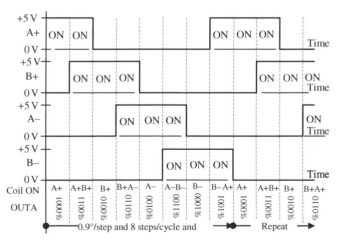

■ **FIG. 4.33** "Half-step" mode voltage waveforms to the transistor gates of the unipolar circuit.

Here is a program that programs the half-step mode:

```
period       VAR   word
startperiod  CON   50000              'be careful that a word variable cannot exceed 65,535
minperiod    CON   5000
again:  FOR period=startperiod TO minperiod        'ramp up from min speed to max speed
        OUTA=%0001          'step 1. There are now 8 steps per cycle in half step mode
        PULSOUT 4,period    'pin 4 is not being used so send out a pulse as a delay
        OUTA=%0011          'step 2
        PULSOUT 4,period
        OUTA=%0010          'step 3
        PULSOUT 4,period
        OUTA=%0110          'step 4
        PULSOUT 4,period
        OUTA=%0100          'step 5
        PULSOUT 4,period
        OUTA+%1100          'step 6
        PULSOUT 4,period
        OUTA=%1000          'step 7
        PULSOUT 4,period
        OUTA=%1001          'step 8
        PULSOUT 4,period
        NEXT

        FOR period=minperiod TO startperiod    'ramp down from max speed to min speed
        OUTA=%0001
        PULSOUT 4,period
        OUTA=%0011
        PULSOUT 4,period
        OUTA=%0010
        PULSOUT 4,period
        OUTA=%0110
        PULSOUT 4,period
        OUTA=%0100
        PULSOUT 4,period
        OUTA+%1100
        PULSOUT 4,period
        OUTA=%1000
        PULSOUT 4,period
        OUTA=%1001
        PULSOUT 4,period
        NEXT
        GOTO again
```

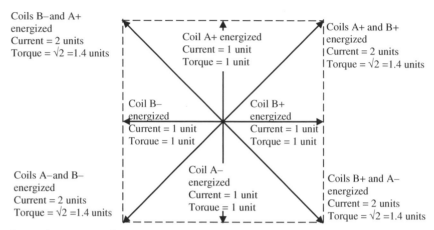

■ **FIG. 4.34** Phase diagram of stepper motor coil energization.

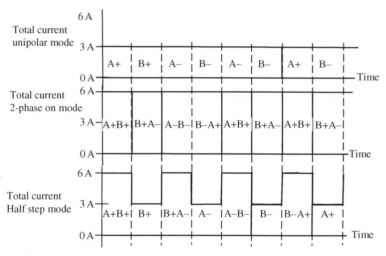

■ **FIG. 4.35** Total current for the three modes of stepper motor driving.

Figs. 4.34 and 4.35 summarize the currents taken and the torques produced, respectively, by the stepper motor.

The Bipolar Circuit

A bipolar circuit is based on the H-bridge, Fig. 4.36. The Keyes bipolar driver in Fig. 4.2 uses the L298 stepper motor drive chip and the reader is encouraged to Google the data sheet for this integrated circuit. The H-bridge converts a unipolar power supply to a bipolar power supply such that the current in each

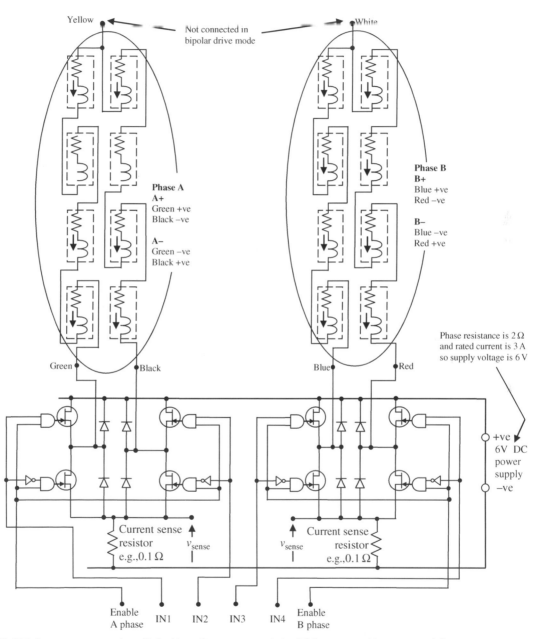

■ **FIG. 4.36** Stepper motor connected to a bipolar driver with a current sense circuit which is not connected because not used. Current sensing is used when chopping control of current is required. Chopping control is required when power supply exceeds 6 V and is done to improve motor performance. Transistor switching sequence is, see truth table below:

Full step	A+,		B+,		A−,		B−
2-Phase-on step		A+B+,		B+A−,		A−B−,		B−A+
Half step	A+,	A+B+,	B+,	B+A−,	A−,	A−B−,	B−,	B−A+

phase can be reversed. Note that in bipolar mode the yellow and white wires are not connected, and "a phase" is a connection between:

(i) green and black and is called **phase A**
(ii) blue and red and is called **phase B**

Furthermore, in phase A

(i) positive current in phase A is annotated, **A+,** that is, green is +ve and black is −ve
(ii) negative current in phase A is annotated, **A−**, that is, green is −ve and black is +ve

…and similarly in phase B

(i) positive current in phase B is annotated, **B+,** that is, blue is +ve and red is −ve
(ii) negative current in phase B is annotated, **B−**, that is, blue is −ve and red is +ve

The bipolar driver will give twice the torque of the unipolar driver because twice the number of coils is used at each step motion of the rotor. A bipolar driver drives current through each coil phase in two directions, hence the name, "bipolar," that is, forward current and reverse current. Fig. 4.34 shows a circuit diagram of the four phases of the stepper motor connected to a bipolar driver. It uses a circuit called an "H-bridge" because it looks like the letter "H."

Truth table (X = do not care)							
Phase	Enable A phase	IN1	IN2	IN3	IN4	Enable B phase	
A+	1		1	0	X	X	0
B+	0		X	X	1	0	1
A−	1		0	1	X	X	0
B−	0		X	X	0	1	1

4.11 **THE L298 "KEYES" BIPOLAR DRIVER**

Fig. 4.37 shows the wiring diagram between the between stepper motor, a Keyes bipolar driver, and a Basic Stamp microcomputer. The Basic Stamp program, below, rotates the motor 1 revolution CW in half-step mode (400 steps/rev), waits for 0.5 s, and then repeats the same procedure. Stepping frequency is 100 Hz due to a period of 10 ms.

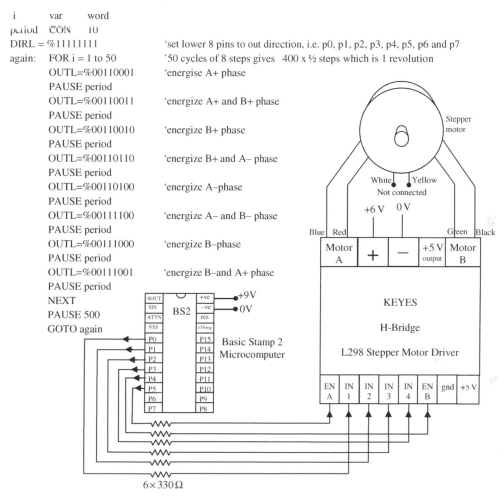

```
i          var    word
period     CON    10
DIRL = %11111111                'set lower 8 pins to out direction, i.e. p0, p1, p2, p3, p4, p5, p6 and p7
again:     FOR i = 1 to 50      '50 cycles of 8 steps gives   400 x ½ steps which is 1 revolution
           OUTL=%00110001       'energise A+ phase
           PAUSE period
           OUTL=%00110011       'energize A+ and B+ phase
           PAUSE period
           OUTL=%00110010       'energize B+ phase
           PAUSE period
           OUTL=%00110110       'energize B+ and A– phase
           PAUSE period
           OUTL=%00110100       'energize A–phase
           PAUSE period
           OUTL=%00111100       'energize A– and B– phase
           PAUSE period
           OUTL=%00111000       'energize B–phase
           PAUSE period
           OUTL=%00111001       'energize B–and A+ phase
           PAUSE period
           NEXT
           PAUSE 500
           GOTO again
```

■ **FIG. 4.37** Wiring diagram between (i) stepper motor, (ii) Keyes bipolar driver, and (iii) Basic Stamp microcomputer.

4.12 LEADSHINE DM542 BIPOLAR MICROSTEPPING CURRENT CHOPPING DRIVER

The Leadshine DM542 stepper motor driver is a whole new ball game. A wonderful device that gives beautiful smooth and quiet drive to the stepper motor when used in microstepping mode. It cannot work in 200 step/rev full step mode and starts at 400 steps/rev half-step mode and can drive the stepper motor in microsteps up to 25,600 steps/rev! Do not forget though that microstepping gives you high resolution but does not give you more accuracy because the load applied to the motor shaft will still shift the phase of the shaft by up to ±0.45 degrees, see Fig. 4.19. The driver gives a discretized 2-phase sinusoidal polyphase current drive to the stepper motor

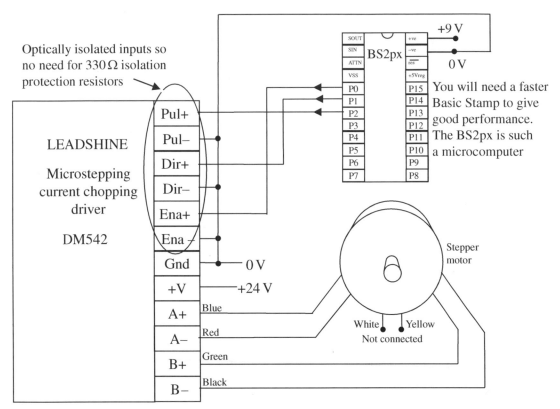

■ **FIG. 4.38** Wiring diagram between (i) stepper motor, (ii) DM542 bipolar microstepping current chopping driver, and (iii) Basic Stamp 2px microcomputer.

transforming it into a very smooth low noise, low vibration, brushless DC, open-loop, synchronous motor. The DM542 can handle a supply voltage that ranges from 20 to 50 V. Furthermore, it has a chopping current circuit that maintains the current at a preselectable level from 1.00 to 4.20 A in 0.46 A increment using switches on the side of the driver box. Fig. 4.38 shows the DM542 driver connected to the stepper motor and a BS2px high-performance Basic Stamp microcomputer that executes instructions faster than lower specification microcomputers in the Basic Stamp range.

One major advantage of the DM542 is that there is no need to switch consecutively the A+, A−, B+, and B− phases. The DM542 does that automatically. All that has to be done is to set the direction of the motor, either CW or CCW, with a logic high or low on the Dir+ pin, Fig. 4.38. Once the direction has been set then the phases are advanced one by one by sending a logic high pulse of minimum length 2.5 μs to pin Pul+. For example, here is a BS2px program that accelerates the motor from a starting minimum speed to a maximum speed then stays at the maximum speed for 1 revolution, then decelerates back to the minimum speed, then comes to rest after 0.1 s, and then switches the current off. The clock period of a BS2px microcomputer is 0.81 μs.

```
enable CON   0          'enable driver with a logic low
dir    CON   1          'logic low is one direction, e.g. CW and logic high is opposite
                          direction, e.g. CCW
pul    CON   2          'high pulse of greater than 2.5µs, i.e. 4 clock pulses (=3.24µs)
a      VAR   Word
i      VAR   Word
       LOW   enable     'activates driver to enable motion
       LOW   1          'set one direction, e.g. CW
       a=2800           'set starting step period, i.e. the starting step frequency
       FOR i = 1 TO 400 'accelerate for one revolution, DM542 already switched to 400steps/rev
       PULSOUT pul,4    'move on by one step
       PULSOUT 3,a      'wait, i.e. delay for the step periodic time, this is a dummy wait pulse
       a=a-6            'decrease the period in order to accelerate the motor
       NEXT

       a=400            'set period to give maximum speed

       FOR i = 1 TO 400 'drive for one revolution at constant maximum speed
       PULSOUT pul,4
       PULSOUT 3,a
       NEXT

       FOR i = 1 TO 400 'decelerate for one revolution
       PULSOUT pul,4    'move on by one step
       PULSOUT 3,a      'wait, i.e. delay for the step periodic time, this is a dummy wait pulse
       a=a+7            'increase the period in order to decelerate the motor
       NEXT

       PAUSE   100      'wait 0.1s
       HIGH enable      'switch off current to the motor by disabling the driver

again: GOTO again       'stop programme with an infinite loop
```

4.13 **PROBLEMS**

1. Analyze the acceleration versus time for the acceleration algorithms
    described in this chapter. These algorithms give acceleration by
    decrementing the periodic time between steps by a constant value.
    Show, via Excel spreadsheet or otherwise, that such algorithms cause
    acceleration to increase with respect to time, which is not a good idea
    since the torque decreases with speed. A better algorithm is to arrange
    the acceleration to decrease with respect to time so as to match the
    decreasing torque that is available. Remember that this is not so easy

because any mathematical computation that computes the next periodic time encroaches on that periodic time and thus both spoils and limits the computation.

2. The Leadshine DM542 driver has a supply voltage range from +20 to +50 V. Try connecting three and four 12 V power supplies in series and see how the performance improves in terms of pull-in and pull-out speed. Make sure you do this under supervision because any erroneous wiring can be unsafe and also lead to permanent damage of the driver. The pull-in speeds are likely to be too fast for the Basic Stamp so you will need to uprate to a Basic Stamp2SX or 2PX. If these are not fast enough you'll need to discuss with your supervisor. These pull-in and pull-out speeds will change when you apply the stepper motor to the Hitter and Thrower robots since the speeds are a function of inertia load.

Phase	Enable A phase	IN1	IN2	IN3	IN4	Enable B phase
A+	1	1	0	X	X	0
B+	0	X	X	1	0	1
A−	1	0	1	X	X	0
B−	0	X	X	0	1	1

Truth table (X=do not care)

Chapter 5

Theory IV: Collision Theory and Design Notes Related to the Hitting Robot

LEARNING OUTCOMES

1. Basic understanding of Newton's laws and the laws of rectilinear collision.
2. Appreciation of force impulse and change of momentum.
3. Understanding the design principles of the Hitting robot mechanical system components.

5.1 INTRODUCTION

Fig. 5.1 shows the basic principle behind the operation of the Hitting robot. The idea is similar to many sports activities that involve hitting a ball, such as golf, cricket, tennis, etc. There are three important engineering aspects to the Hitting robot that are:

(i) The theory of collision dynamics.
(ii) How to design an efficient ball Hitting robot, knowing the theory of collision dynamics.
(iii) How to design an accurate ball Hitting robot, that is, how can we engineer an effective robot?

We address these aspects by

(i) first defining Newton's laws of rectilinear collision,
(ii) ...then to measure the coefficient of restitution between the ball and the Hitting pad,
(iii) ...then to describe the robot engineering design, and
(iv) ...finally, to explain what happens during collision.

5.2 THEORY OF COLLISION DYNAMICS

Newton's laws of collision for rectilinear (straight line) motion between two bodies are as follows:

Creating Precision Robots. https://doi.org/10.1016/B978-0-12-815758-9.00005-5

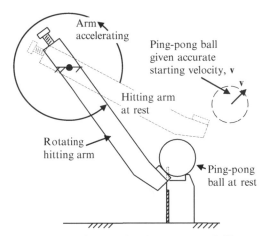

Arm accelerating

Ping-pong ball given accurate starting velocity, **v**

Hitting arm at rest

Rotating hitting arm

Ping-pong ball at rest

■ **FIG. 5.1** Diagram showing rotating hitting arm that is hitting a ping-pong ball from rest to give an accurate velocity, **v** to the ball.

1. Newton's Collision Law 1

The ratio of the relative speed after collision to the relative speed before collision is equal to the coefficient of restitution, u, where u varies between 0 for perfectly inelastic objects and 1 for perfectly elastic objects.

2. Newton's Collision Law 2

Total momentum is conserved, that is, total momentum before collision is equal to total momentum after collision. However, total kinetic energy (KE) may not be conserved but if $u = 1$ then total KE is conserved but if $u < 1$ then some of the total KE is lost that is usually manifested as raising of the temperature of the bodies and/or as work done in the plastic deformation of one or both of the bodies. This can also be thought of as hysteretic work.

5.3 MEASUREMENT OF THE COEFFICIENT OF RESTITUTION, *U*

We will address the collision law 1 by investigating the dynamical behavior of a ping-pong ball by measuring its coefficient of restitution when impacting a similar, replicated surface of the hitting pad of the Hitting robot.

The coefficient of restitution, u can be found by a bounce test according to Fig. 5.2. Here, a ping-pong ball is dropped onto a 4 mm thick piece of acrylic, that is, Perspex, that is double-sided sticky taped to a hard and smooth floor. Acrylic is chosen because this material will be used for the Hitting robot hitting pad. Next, drop the ball alongside a vertical standing

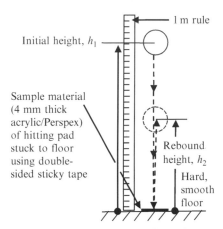

1 m rule

Initial height, h_1

Sample material
(4 mm thick
acrylic/Perspex)
of hitting pad
stuck to floor
using double-
sided sticky tape

Rebound
height, h_2

Hard,
smooth
floor

■ **FIG. 5.2** Drop ball bounce experiment to determine its coefficient of restitution, *u*.

1 m rule and get a class mate to help you; one of you to hold the rule and drop the ball from height h_1 and the other to step back and to read the bounce height h_2. Try to be accurate with your measurements to the nearest 1 cm.

A bounce experiment carried out according to Fig. 5.1 gave the values, $h_1 = 1.00$ m and $h_2 = 0.65$ m.

Using Newton's collision law 2, and assuming negligible air drag resistance, the approximate value of the coefficient of restitution, *u* is given by the following reasoning:

$$\text{Coefficient of restitution, } u = \frac{\text{relative speed after collision}}{\text{relative speed before collision}}$$

Now recall the equations of rectilinear motion for a body under constant acceleration:

(i) $v = u + at$
(ii) $v^2 = u^2 + 2as$
(iii) $s = ut + \frac{1}{2}at^2$

...where, t = time, u = initial speed, v = final speed, a = acceleration, and s = distance traveled in time t.

Using Eq. (ii) above, the final speed, v_1 of a dropped ball, starting at rest under gravitational acceleration, g, from a height, h_1, is given by, $v_1 = \sqrt{(2gh_1)}$. Conversely, the speed, v_2 of a ball immediately after rebounding from a collision that rises to a height, h_2, is given by, $v_2 = \sqrt{(2gh_2)}$, thus, coefficient of restitution, *u* is given by

$$u = \frac{\text{relative speed after collision}}{\text{relative speed before collision}} = \frac{\sqrt{2gh_2}}{\sqrt{2gh_1}}$$

$$\text{coefficient of restitution, } u = \sqrt{\frac{h_2}{h_1}} \qquad (5.1)$$

$$\text{Hence, coefficient of restitution, } u = \sqrt{\frac{0.65}{1.00}} = \underline{0.81} \qquad (5.2)$$

Note that more than two significant figures for the value of, u is not warranted because the best precision for measuring h_1 and h_2 is to an error of 0.01 m, that is, 1 cm, and such precision leads to an error of 0.01 in the computation of, u (e.g., try $h_1 = 0.65 \pm 0.01$ m and $h_2 = 1.00 \pm 0.01$ m), thus three significant figures (in this case 3 dec.pl.) computation is an unjustified calculation. This calculation methodology is important and the authors note that students nowadays need to be reminded of this important engineering discipline concerning tolerancing, errors, and precision.

The authors proposed to the student class to use cardboard for the hitting pad, that is, the surface that comes into contact with the ball during collision. Later, a student said he had done tests replacing the cardboard with acrylic (Perspex) for the hitting plate and the hitting distance had significantly improved. This suggestion is what led the authors to devise the ball bounce test in order to prove the superiority of Perspex over cardboard in enhancing the coefficient of restitution. Students are encouraged to carry out tests using different material surfaces but do not forget that it is also the material and system dynamics behind the hitting plate that affects the coefficient of restitution.

Before we introduce discussion of Newton's collision law 2 in order to explain what happens during the impact between the ball and the hitting surface, it is necessary to describe the constructional design features of the Hitting robot.

5.4 **HITTING CONSTRUCTIONAL DESIGN FEATURES**

The Hitting robot is a mechatronic system that uses a microcomputer controlled stepper motor to rotate directly a hitting arm. At the extremity of the arm is mounted a hitting pad. The pad strikes, and simultaneously imparts a precise force impulse to the ping-pong ball such that the ball is given an accurate launching velocity. The idea is to hit the ping-pong ball into waste paper baskets located at 2, 4, and 6 m range. In a highly simplified form, the Hitting robot hits the ping-pong ball much like a golfer swings a golf club to strike a golf ball at "tee off." The golf ball is supported a few centimeters above the turf with a "tee." The tee allows the swinging club to strike the ball below its horizontal center thus permitting an initial upward launching angle of approximately 20 degrees from the horizontal. The situation is more complicated since the golf ball will be given backspin that will cause the ball to "fly" using the Magnus effect. Nonetheless, such an angle turns out to be the optimum launch angle for a golf ball to achieve maximum range but for the ping-pong ball in our application, largely without spin, the optimum angle is 40 degrees. This is an interesting research area for students where they can consider redesigning their Hitting robot to emulate a golf club to give backspin.

The key features of the Hitting robot for striking the ping-pong ball accurately are:

(i) A rigidly mounted stepper motor that can be driven at an accurately controlled angular speed.
(ii) A tee for precisely positioning the ping-pong ball without impeding its launching velocity.
(iii) A rotating hitting arm of sufficient length with a hitting pad at its extremity.

Let us address each feature in turn.

First of all, the Hitting robot has a rigid base and rigidly mounted to it is a stiff tower that holds rigidly at its top the stepper motor. So, this means that any stepper motor vibration amplitude will be low and positioning repeatability will be high. Next, the stepper motor is driven by microcomputer impulses that are derived from an onboard resistor-capacitor (RC) oscillator clock that produces a timed set of pulses for driving the stepper motor. The stepper motor is a highly accurate actuator that when driven by an accurately timed set of pulses will give a very accurate rotation speed. Any inaccuracy in angular velocity can be blamed on the timed set of pulses; not on the stepper motor. So, attention should be paid to the RC clock frequency which is subject to temperature variation. In the first instance, students should get results by hitting balls into baskets. Later, they should address repeatability errors. One such repeatability error is the microcomputer clock frequency.

Second, we will design a tee with a three-point mount for the ball. A three-point mount gives the minimum number of points for a highly repeatable positioning methodology. If there are more than four points, then one point may or may not be supporting the ball with a result that the ball positioning repeatability will be poor and as a result the launching velocity will not be repeatable. The tee must also be designed such that the hitting arm can pass through or pass by the tee, and in the process striking the ball, without touching the tee or itself being impeded by the tee. Also the ball, on being struck, should be lifted straight off its three-point mounting.

Third, the hitting arm is dimensioned such as to place the hitting pad at a radius of 0.26 m. The angular speed of the stepper motor was estimated to be 5 revs/s and the increase of ball speed over its hit speed (the hitting pad speed) was estimated to be 50%. In order to achieve 6 m range, the launching speed was estimated to be 12 m/s (Chapter 3), hence the hitting pad speed $= 2/3 \times 12 = 8$ m/s and the angular speed of the hitting arm is $\omega = v/r = 8/0.26 = 31$ rad/s which is approximately 5 rev/s. We do not know yet if a stepper motor speed of 5 rev/s is possible but if it is not then we either compromise with a lower hitting range or we rebuild the robot with a longer arm. This is the art and science of prototype product design where you have to deal with change and compromise. Interestingly, if the length of the arm is increased, then so also will the polar second moment of inertia of the arm increase. This means that it may not be possible to use the available torque from the stepper motor to accelerate the arm in less than one revolution to 5 rev/s. This maximum available rotation angle is all that is allowed for accelerating the arm since the ball is resting in the tee awaiting collision.

Ok, with these features in mind, let us set about designing the tee and the hitting arm. The stepper motor rigid mounting will be taken care of in its construction that is described in Chapter 10.

5.5 DESIGN OF THE TEE

Fig. 5.3 illustrates (i) the rotating hitting arm extremity on which is mounted the hitting pad, (ii) the tee, and (iii) the ping-pong ball. A U-shaped slot cut into the tee allows the rotating hitting arm to swing through the tee without impediment. Furthermore, the slot allows the swinging arm to strike the ball at a point, p that is 40 degrees from the horizontal. This angle has been chosen because, arguably, it has been found to give the maximum ball hitting range.

Note that the information shown in Fig. 5.3 legend box. In particular, note that the three-point-mounting arrangement at points a, b, and c and note that the arcs e_1-e_2 and e_3-e_4 pass through the U-shaped slot unimpeded with clearance. A side view of the tee, Fig. 5.4, is highly informative because it shows a side view of the arm and the tee where it can be seen how the ball is perched on the three-point mounting and at the same time the construction of the tee allows the hitting pad to pass unimpeded. Furthermore, from Fig. 5.4, it can be seen that the angle ϕ which is subtended by points, *hmc* where line, *hm* is vertical, must be less than the launch angle of

Ping-pong ball
Rotating direction of arm
Hitting pad
Extremity of hitting arm
Ball supporting tee
U-shaped slot in tee allows the hitting to pass through unimpeded
m
a
b
c
p
d
f
e1
e2
e3
e4
Legend
a, b, c Tee three-point mounting of ball
d, p, f Locus of hitting pad contact point between pad and ball
e1 to e2 Loci of outer edge
e3 to e4 points of hitting pad showing how the loci pass uninterrupted through the tee U-shaped slot
m Ball center of mass

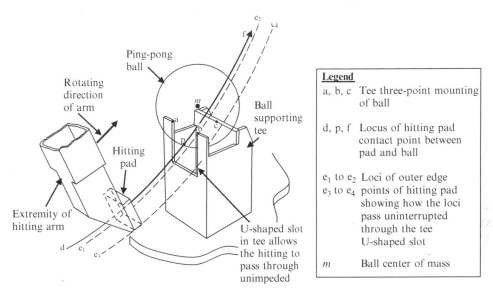

■ **FIG. 5.3** The hitting arm shown passing unimpeded through the tee. Note that the hitting pad, when colliding with the ball, cleanly lifts the ball off the three-point mounting points of the tee to give a highly repeatable launching velocity.

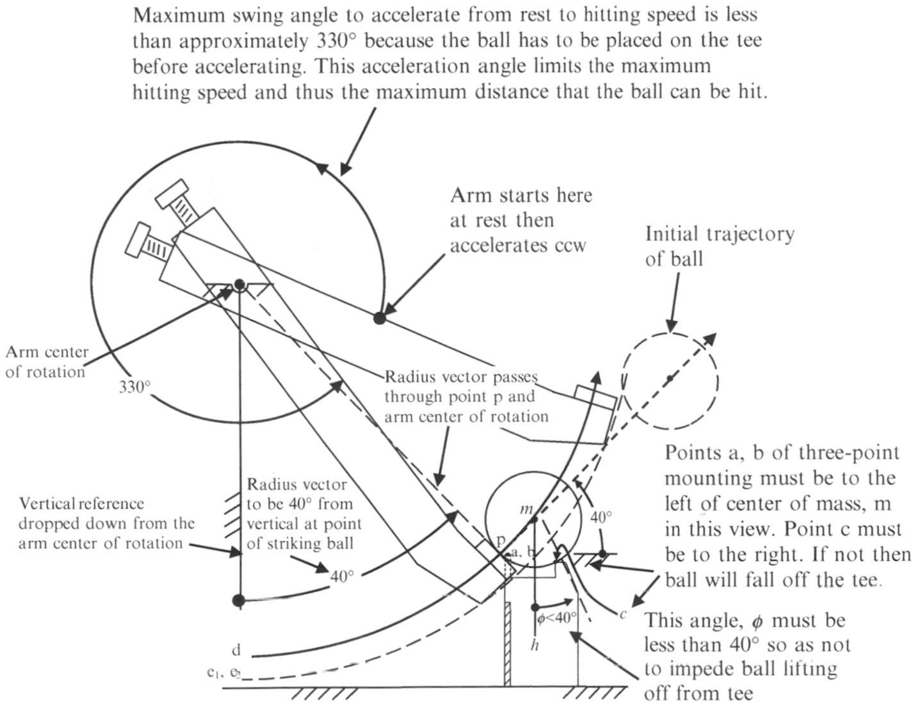

Maximum swing angle to accelerate from rest to hitting speed is less than approximately 330° because the ball has to be placed on the tee before accelerating. This acceleration angle limits the maximum hitting speed and thus the maximum distance that the ball can be hit.
Arm starts here at rest then accelerates ccw
Initial trajectory of ball
Arm center of rotation
330°
Radius vector passes through point p and arm center of rotation
Points a, b of three-point mounting must be to the left of center of mass, m in this view. Point c must be to the right. If not then ball will fall off the tee.
Vertical reference dropped down from the arm center of rotation
Radius vector to be 40° from vertical at point of striking ball
40°
m
p
a, b
c
40°
ϕ<40°
h
d
e1, e2
This angle, ϕ must be less than 40° so as not to impede ball lifting off from tee

■ **FIG. 5.4** Diagram showing (i) hitting arm limited swing angle for acceleration to hitting speed and (ii) a side view of the ball tee and some of its geometrical design features.

40 degrees so that the ball can be lifted cleanly, without impediment, off its three-point mounting. If angle φ is more than 40 degrees, then the ball will be pushed into the tee causing permanent damage and badly spoiling the launching velocity. Also, note that the tee should be placed such that the hitting pad contacts the ball at point, p just as the radius vector becomes 40 degrees from the vertical. Fig. 5.4 also shows that only 330 degrees is available for the arm to swing from rest to its full hitting speed. Thus, the arm should be as light as possible in order for the stepper motor to use its torque to accelerate the arm at a high value in order to attain a high angular velocity in as short an angle as possible. Note that in Fig. 5.4, points a, b of the three-point mounting are to the left of the ball center of mass, *m* otherwise the ball will fall off the tee. In other words, a stable support arrangement is when the ball weight vector lies inside the ball support polygon.

5.6 DESIGN OF THE HITTING ARM

There are some constructional design features and dynamical characteristics to be described concerning the hitting arm as shown in Fig. 5.5. These are listed as follows:

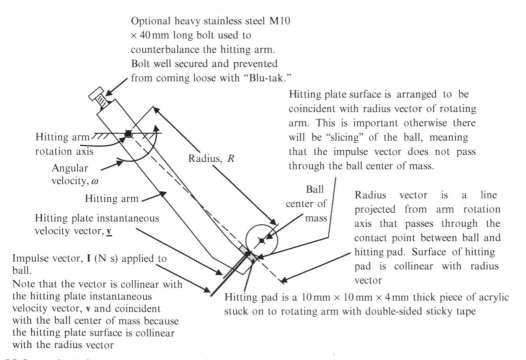

Optional heavy stainless steel M10 × 40 mm long bolt used to counterbalance the hitting arm. Bolt well secured and prevented from coming loose with "Blu-tak."

Hitting arm rotation axis

Angular velocity, *ω*

Hitting arm

Radius, *R*

Hitting plate instantaneous velocity vector, **v**

Impulse vector, **I** (N s) applied to ball.
Note that the vector is collinear with the hitting plate instantaneous velocity vector, **v** and coincident with the ball center of mass because the hitting plate surface is collinear with the radius vector

Hitting plate surface is arranged to be coincident with radius vector of rotating arm. This is important otherwise there will be "slicing" of the ball, meaning that the impulse vector does not pass through the ball center of mass.

Ball center of mass

Radius vector is a line projected from arm rotation axis that passes through the contact point between ball and hitting pad. Surface of hitting pad is collinear with radius vector

Hitting pad is a 10 mm × 10 mm × 4 mm thick piece of acrylic stuck on to rotating arm with double-sided sticky tape

■ **FIG. 5.5** Diagram showing hitting arm constructional design features and the dynamical characteristics during collision between the hitting plate and the ball.

1. A relatively heavy stainless steel bolt is used to balance the arm about its rotation axis. This has the advantage of almost eliminating rotating arm out-of-balance forces which thus leads to a more stable structure when the arm is rotating that leads to a more accurate apparatus. The disadvantage is that the polar moment of inertia, J is increased which reduces the arm angular acceleration capability; recall that stepper motor torque, $T = J \cdot \alpha$, where α is the angular acceleration of the arm. Thus, angular acceleration, $\alpha = T/J$ is reduced since J has been increased with the added mass. Angular acceleration is important because the hitting arm has less than 1 revolution to increase its angular velocity from rest to a suitable hitting velocity, Fig. 5.5. It is considered that the reduction in angular acceleration capability in order to increase target accuracy was a better choice.
2. It is most important that the hitting arm radius vector is collinear with the hitting plate surface. If not then collision complications set in leading to inaccuracy of hitting the target.
3. Now that the radius vector is coincident with the hitting plate surface, the impulse vector I is coincident with (i) the instantaneous velocity vector of the point of contact between the hitting plate and the ball and (ii) the ball center of mass. If not there will be the possibility of "David Beckham spin." In fact such complications can be used as an interesting ball dynamics challenge for an advanced course where spin can be created in a horizontal axis or combined with a spin component in the vertical axis.
4. If the robot and the tee are connected as part of the same structure, then the stepper motor is likely to induce troublesome vibrations into the tee and this will cause the ball to jump around on its mounting points. Vibration creation is one of the disadvantages of the stepper motor due to the sudden jerk, or shock, brought about when stepping between phases. The jerks are the worst in full-step mode and are reduced in half-step mode are further reduced to a very low value in microstepping mode. To reduce and absorb the stepping jerks, robots are mounted on 1 cm diameter hemispherical stick-on rubber feet.

With the construction description of the rotating hitting arm and the tee completed, we now move on to describe the physics that occurs during the hitting pad colliding with the ping-pong ball.

5.7 **ANALYSIS OF BALL COLLISION**

With reference to Fig. 5.6, which shows the stages of the ball-hitting collision, Newton's two laws of collision are now used to analyze the collision dynamics of hitting the ball, as follows.

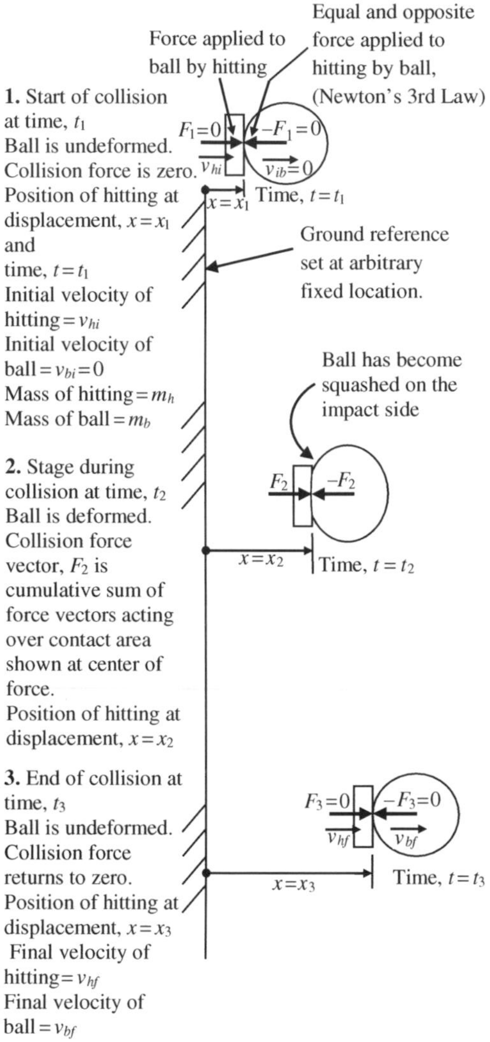

FIG. 5.6 Three stages of collision between the hitting and the ball. Note that deflection of the hitting is assumed to be negligible.

First of all, the following nomenclature is used:

1. Mass of hitting $= m_h = 0.135\,\text{kg}$

(Estimated, taking $^2/_3$ of mass of hitting arm because it is a rotating body and we are assuming rectilinear motion)

2. Mass of ball $= m_b = 0.0027\,\text{kg}$

3. Hitting initial velocity before collision $= v_{hi}$

(v_{hi} is known and depends on the programed angular velocity of the hitting arm)

4. Hitting final velocity after collision $= v_{hf}$

(Unknown value)

5. Ball initial velocity before collision $= v_{bi} = 0$

(Ball is at rest)

6. Ball final velocity after collision $= v_{bf}$

(Unknown value and is fundamental to calculate)

7. Hitting/ball coefficient of restitution $= u$
8. Ball-to-hitting mass ratio $= m_b/m_h = r$

Now use Newton's collision laws.

Newton's Collision Law 1 (velocities to the right are positive)

$$\frac{\text{Final relative velocity}}{\text{Initial relative velocity}} = \frac{v_{bf} - v_{hf}}{v_{hi} - v_{bi}^{\,0}} = u$$

$$\text{Thus,} \quad v_{bf} = u\, v_{hi} + v_{hf} \qquad (5.3)$$

Note that if there is no change in velocity of the heavier hitting, that is, $v_{hf} = v_{hi}$ and, the coefficient of restitution, $u = 1$, then the ball velocity, v_{bf} is twice the initial velocity of the hitting! that is, $v_{bf} = 2v_{hi}$

Newton's Collision Law 2

Total initial momentum $=$ total final momentum

$$m_b v_{bi}^{\,0} + m_h v_{hi} = m_b v_{bf} + m_h v_{hf}$$

now divide through by m_h

$$v_{hi} = \frac{m_b}{m_h} v_{bf} + v_{hf} \qquad (5.4)$$

note that ball-to-hitting mass ratio, $r = \frac{m_b}{m_h}$

$$\text{Eq. (5.1)} - \text{Eq. (5.2)} \implies v_{bf} - v_{hi} = u v_{hi} - r v_{bf}$$

$$\text{Thus, final ball velocity, } v_{bf} = v_{hi}\frac{(1+u)}{(1+r)} \tag{5.5}$$

Substituting into Eq. (5.3) the following values:

$$u = 0.81$$

$$\left.\begin{array}{l} m_b = 0.0027 \text{ kg} \\ m_h = 0.0027 \text{ kg} \end{array}\right\} r = \frac{m_b}{m_h} = \frac{0.0027 \text{ kg}}{0.135 \text{ kg}} = 0.020$$

Thus, using Eq. (5.3), the value of v_{bf} is given by

$$v_{bf} = v_{hi}\frac{(1+0.81)}{(1+0.020)} = 1.76 v_{hi} \tag{5.6}$$

Hence, Eq. (5.6) shows an important advantage of the hitting, the advantage being that the ball is launched at a significantly higher velocity than that of the hitting pad velocity, that is, 76% faster. In fact, as the relative ball-to-hitting mass ratio decreases to zero and the coefficient of restitution increases to 1, the ball launch velocity becomes twice that of the hitting velocity. This is a theoretical maximum. However, the disadvantage of the hitting is that the launch velocity is a function of the coefficient of restitution and the ball-hitting mass ratio. Both these values are not easily measured but nonetheless can be calibrated. Even then the coefficient of restitution may be a function of temperature and hitting pad initial velocity and may even vary with different balls. Hence, the Hitting robot presents some interesting research areas for students to explore.

A more detailed analysis of the collision should include rotation, θ_{rot}, of the hitting arm during contact with the ball, Fig. 5.7. Such an effect is likely to cause ccw back spin due to the offset force applied to the ball.

It is interesting to evaluate what happens during the ball-hitting collision. Fig. 5.8 shows a speculative force-time graph that ties up with the three stages of collision shown in Fig. 5.6.

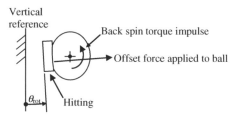

■ **FIG. 5.7** More accurate modeling of the collision.

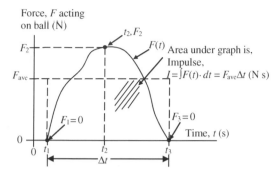

■ **FIG. 5.8** Force-time relationship on ball during collision.

Note that Fig. 5.8 simplifies the irregularity of the $F(t)$ relationship into an average force, F_{ave} lasting for the duration, Δt of the collision where $I = \int F(t) \cdot dt = F_{ave}\Delta t$. We do not have knowledge of $F(t)$ but the impulse, $I = \int F(t) \cdot dt$ can be evaluated from the product of the ball mass and the launch velocity of the ball immediately at the end of the collision. However, the launch velocity cannot be measured directly since we have no means of measuring it. We only can evaluate the launch velocity indirectly by calling upon, (i) the Excel spreadsheet simulation in conjunction with experimental data of the ball flight range plus the ball launch angle and/or (ii) knowing the stepper motor programed angular velocity, ω of the hitting arm together with the arm radius, R, and the coefficient of restitution and the ball-arm mass ratio, the launch velocity can be calculated followed by calculation of ball change in momentum which is equal to the impulse, I.

So, for example, if the programed arm angular velocity is 30 rad/s; the arm radius to the point of impact with the ball is 0.26 m and the velocity gain is 1.76 (see Eq. 5.4) then the launch velocity is 30 rad/s × 0.26 m × 1.76 = 13.7 m/s. Thus, the ball change in momentum is 13.7 m/s × 0.0027 kg = *0.0370 N s*.

Now, an interesting point is reached. If we know the duration Δt of the collision, then F_{ave} can be calculated. But Δt is also an unknown value since a high-speed camera to measure it is not available. What can be done is to speculate for values of Δt, for example, if $\Delta t = 1$ ms, then $F_{ave} = 0.0370$ N s/0.001 s = 37.0 N, that is 3.7 kg f! ...a large reaction force from a ping-pong ball that is less than 3 g mass! Furthermore, the maximum force will be greater than F_{ave} since F_2 is greater than F_{ave}. If, however, $\Delta t = 10$ ms then F_{ave} is lowered 10 times to 0.37 kg f. If F_{ave} is equal to the larger force of 37.0 N, then the load torque on the arm is 37.0 N × 0.26 m = 9.6 N m and this is enough to stall the stepper motor which has an estimated working

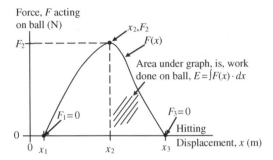

■ **FIG. 5.9** Force-displacement on ball during collision.

torque of less than 1 N m. In fact, this is what happens at high angular velocities, that is, the stepper motor will be stalled, and will not recover its rotation. This happens when the stepper motor is running at an angular velocity greater than its pull-in speed. In contrast, a brushed DC motor will recover due to its electromechanical commutation.

Discussion on the collision dynamics is brought to an end with a final academic note related to the force-displacement relationship acting on the ball. Fig. 5.9 shows a speculative force-displacement, F-x, graph that also ties up with the three stages of collision shown in Fig. 5.6.

The force-displacement relationship in Fig. 5.9 is interesting because it shows the work done on the ball, $E = \int F \cdot dx$ from x_1 to x_3. If the coefficient of restitution is equal to 1, then E will be equal to the final KE of the ball, that is, $\frac{1}{2} m_b v_{bf}^2$. So that the final velocity of the ball, $v_{bf} = \sqrt{(2E/m_b)}$. This result will agree with the momentum solution, Eq. (5.5) with $u = 1$. If, however, the coefficient of restitution is less than 1, then $v_{bf} < \sqrt{(2E/m_b)}$ because not all the work done on the ball goes into increasing its velocity.

The relationships in Figs. 5.8 and 5.9 are speculative because ball forces, ball displacements, and time are not measured and furthermore would require sophisticated experimental apparatus to measure them. This exemplifies Isaac Newton's genius because he deduced these laws approximately 350 years ago, certainly based on experiments but, without sophisticated scientific instruments.

A more repeatable, deterministic, and accurate method of propelling a ball may be to throw it rather than hit it and so the idea of the Thrower robot was born.

The disadvantage however of the Thrower, is that the launch velocity must be 76% faster than the hitting and that leads to the requirement of a higher-performance apparatus.

6

The Ball Hitting Robot: Design and Construction

LEARNING OUTCOMES

1. Experience in building a multisegmented strong, stiff, and lightweight structure and mechanism.
2. How to program a stepper motor for starting speed and acceleration.
3. How to datum a stepper motor.
4. Programming the precision hitting of a ball into a basket at a given range.

6.1 MAKING THE BASE PLATE

This stage will show you how to make a lightweight, stiff, and strong base for the hitting. The principle of its design is similar to the design of aircraft floors.

Note that 1.5 mm thick quality cardboard is used throughout the construction of this robot. The quality cardboard we used was colored black but if you can obtain different colors then your robot will become more attractive. However, remember that the design drawings are for 1.5 mm thick cardboard.

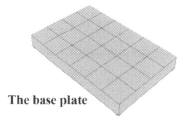

The base plate

First Make the "Instep Jig"

The jig is used to place cardboard pieces at 1.5 mm distance inboard on a cardboard plate. You will understand the significance of this jig after a few steps.

Creating Precision Robots. https://doi.org/10.1016/B978-0-12-815758-9.00006-7

1. Cut one piece $20\,\text{mm} \times 50\,\text{mm}$
2. Cut this piece down the middle giving two pieces $10\,\text{mm} \times 50\,\text{mm}$
3. Glue the two pieces together using a scrap piece of 1.5 mm cardboard to give a step displacement of 1.5 mm as shown

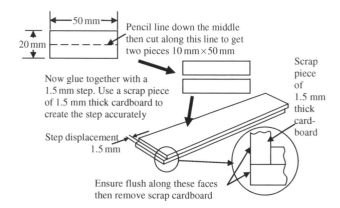

Now Start Making the Parts

Note All Units Are Millimeter and All Parts Are 1.5 mm Thick Cardboard

Remember to cut accurately; aim for $\pm 0.2\,\text{mm}$ error. Check with a measuring caliper. Challenge yourself and make it fun. Remake the part if it is poor quality. The authors work to this accuracy. Sometimes the error goes up to $\pm 0.5\,\text{mm}$ but that should be your absolute maximum error.

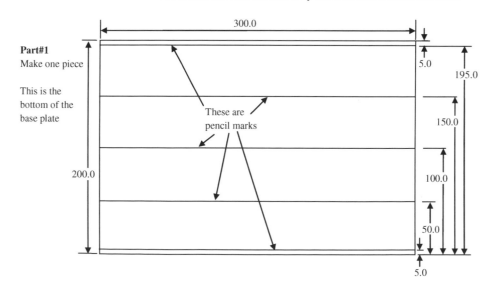

Part#2
Make 30 pieces

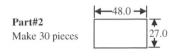

Part#3
Make seven pieces

Part#4
Make two pieces

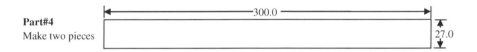

Part#5
Make one piece

This is the top of
the base plate

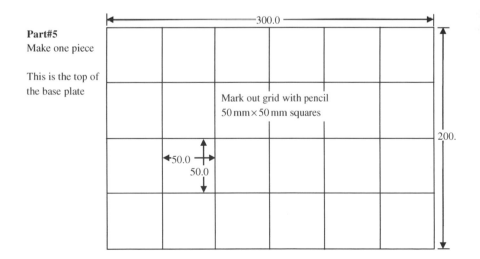

Mark out grid with pencil
50 mm × 50 mm squares

Now assemble these parts by gluing as follows.

Five pieces part#2

Part#1

<u>Glue step#1</u>

Five pieces to be glued as vertical as you can judge and to be glued on the pencil lines and to be glued 1.5 mm in board from 200 mm edge using the instep jig

200 mm edge

<u>The Instep Jig.</u>

Remember do not glue it in place. IT IS A JIG! to be removed after positioning part#2 in place by 1.5 mm in from the edge. Keep the jig clean by regularly wiping it with a tissue. Do not let glue build up on it.

<u>Instep jig</u>

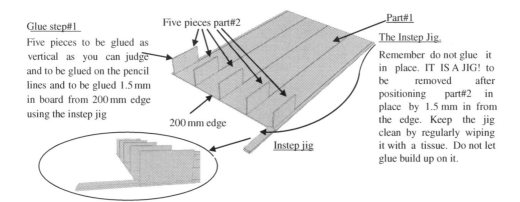

Five pieces part#2

Part#1

<u>Glue step#2</u>

Glue part#3 to both part#1 and all five pieces of part#2 but make sure part#3 is 1.5 mm inboard using the instep jig.

<u>Part#3</u>

<u>Instep jig</u>

<u>Glue step#3</u>

Now glue another part#3 to part#1 so as to glue up against the five pieces of part#2 but make sure you use the instep jig again.

<u>Part#3</u>

<u>Part#1</u>

<u>Part#2</u>
Five pieces

<u>Instep jig</u>

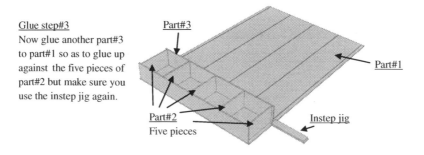

Glue step#4

Now glue another set of five pieces of part#2 onto the pencil lines of part#1 and make sure you glue up against part#3. Once again keep these parts as upright as you can judge. No need for the instep jig for this glue step

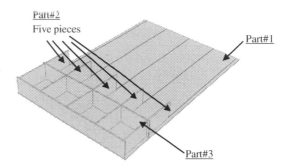

Part#2
Five pieces

Part#1

Part#3

Glue step#5

Now glue another part#3 using the instep jig

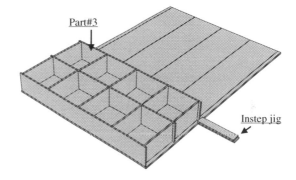

Part#3

Instep jig

Glue step#6

Carry on gluing part#3 and five pieces of part#2 but do not complete the last set because the gluing technique is different. Just complete as far as is shown in the diagram here. Remember to keep using the instep jig for each part#3

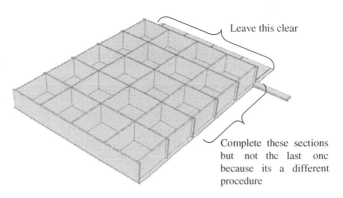

Leave this clear

Complete these sections but not the last one because its a different procedure

Glue step#7

Turn the assembly around so the unfinished end is facing you like in the diagram. Now glue five pieces of part#2 on the pencil lines of part#1 *but do not glue to part#3*. Instead use the instep jig to place the part#2 pieces 1.5 mm inboard of the 200 mm edge of part#1. You should have a gap between the last set of part#2 pieces and the previous part#3. Do not worry that is designed to account for tolerances. If you have cut accurately then the gap will be about 1.5 mm. You may have to trim this last set of part#2 pieces to accommodate the 1.5 mm instep.

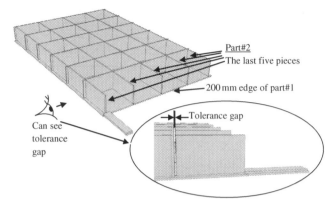

Glue step#8

Now glue the 7th, which is the remaining part#3 into place using the instep jig

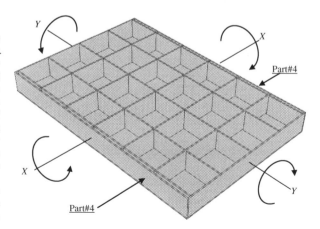

Glue step#9

Now glue the two long sides in place, i.e., part#4 pieces. These pieces should fit neatly into the 1.5 mm insteps created by the instep jig. The top surfaces of all the glued parts should be the same height of 27 mm. If they are not then your base will have a lumpy top surface and will give you trouble as you complete the robot. Its the same as if the foundation of a house that you are building is not good quality. So make sure you cut and glue with accuracy.

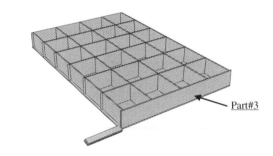

Stiffness and Strength test before gluing top plate to the base

Now do a stiffness and strength test after waiting 15 min after gluing parts#4. Grip opposite ends of the base by two hands and gently try to bend the base about X-X and Y-Y axes but do not try to break it. Notice that it feels stiff and strong in bending. Now try twisting around the X-X and Y-Y axes, do not break it...just a few degrees will do. This time the base feels neither stiff nor strong. Students should appreciate the difference between stiffness and strength...they are completely different. So the base may be strong but it is compliant in torsion and compliance is not good; we want a torsionally stiff base. Let us see if gluing the final part in place solves the problem.

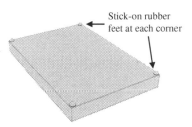

Part#5, the top plate. Make sure you keep the pencil grid lines facing upwards

Glue step#10

Penultimately, glue the last part in place, part#5, i.e. the top plate. Make sure you put glue on all the top edge surfaces of parts #2, #3, and #4 and make sure that you are working on a flat surface. *If you glue the top plate on with the base twisted then it will stay twisted* and you cannot "undo" that. You need to work fast and its a good idea to enlist the help of your classmates. Furthermore, if you can find one, place a heavy flat weight on the top plate to keep the glue in contact with mating surfaces for 15 min. After this time, remove the weight and gently do the previous test for stiffness and strength.

The completed base plate

There will be a remarkable transformation; the base plate will now feel stiff and strong in torsion and even stronger and stiffer in bending. Students should discuss this amongst themselves and with their professor how this stiffness and strength occurs. You will find that the answers tie up with the theory in terms of Mohr's circle, bending of beams, torsional stress, principal stresses and strains. Remember theory does not come from textbooks; it comes from practical experiments

Finally, add four rubber stick-on feet at each corner of the base plate. You can get these from Parallax.com but also from other suppliers. These feet are necessary for two purposes which are, (i) to absorb stepper motor vibration and (ii) to try and inhibit the robot moving around the table that causes robot ball on-target shooting errors.

Stick-on rubber feet at each corner

6.2 **MAKING THE TOWER THAT SUPPORTS THE HEAVY STEPPER MOTOR**

Okay, let us start on making the tower parts.

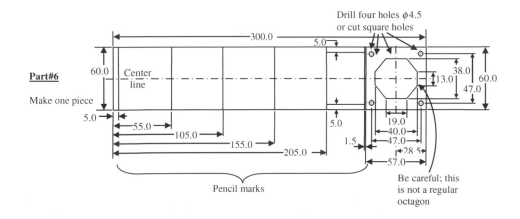

Part#7

Cut one piece
88.0×88.0

Mark out 16 squares of size 22.0×22.0 then mark out diagonals as shown. Then cut into triangles

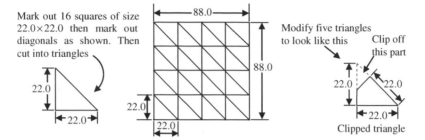

Modify five triangles to look like this

Clip off this part

Clipped triangle

Part#8

Make one piece

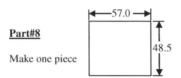

Part#9

Make two pieces

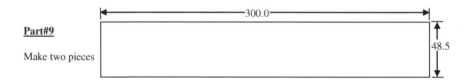

Part#10

Make one piece

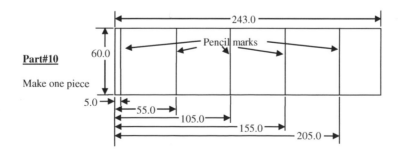

Pencil marks

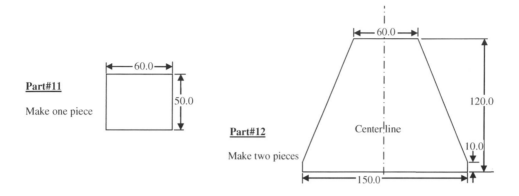

Part#11

Make one piece

60.0

50.0

Part#12

Make two pieces

60.0

120.0

10.0

Center line

150.0

Now Start Gluing Up the Tower

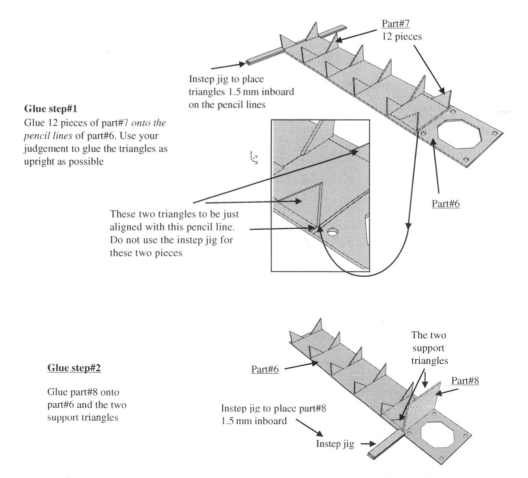

Glue step#1

Glue 12 pieces of part#7 *onto the pencil lines* of part#6. Use your judgement to glue the triangles as upright as possible

Part#7
12 pieces

Instep jig to place triangles 1.5 mm inboard on the pencil lines

These two triangles to be just aligned with this pencil line. Do not use the instep jig for these two pieces

Part#6

Glue step#2

Glue part#8 onto part#6 and the two support triangles

Part#6

The two support triangles

Part#8

Instep jig to place part#8 1.5 mm inboard

Instep jig

Five triangles

Glue step#3

Glue part#9 onto
part#6 and the five
triangles and part#8

Part#9

Part#8

Part#6

Be careful to make
these parts flush, i.e.,
part#6 and part#9

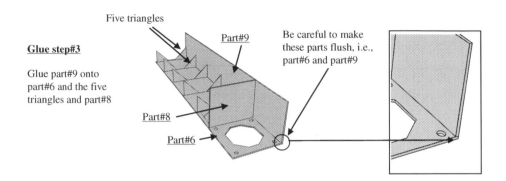

Glue step#4

Repeat gluing part#9
on the opposite side

Part#9

Should be flush at
these two points

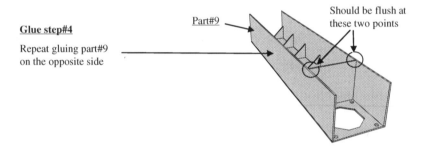

Glue step#5

Glue part#7
triangles onto
the pencil lines
on part#10

Use instep jig to
place the triangles
inboard by 1.5 mm

Part#7
Five clipped
triangles glued on
the pencil lines and
inboard by 1.5 mm

Part#10

Part#7
Five triangles glued on
the pencil lines and
inboard by 1.5 mm

Temporarily put the
partly built tower aside

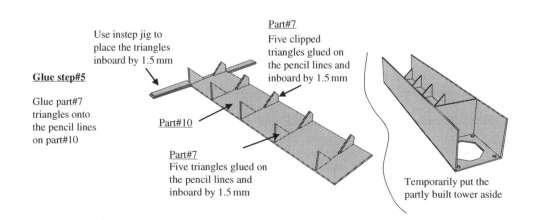

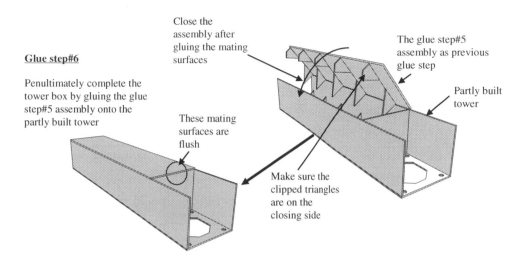

Glue step#6

Penultimately complete the tower box by gluing the glue step#5 assembly onto the partly built tower

Close the assembly after gluing the mating surfaces

The glue step#5 assembly as previous glue step

Partly built tower

These mating surfaces are flush

Make sure the clipped triangles are on the closing side

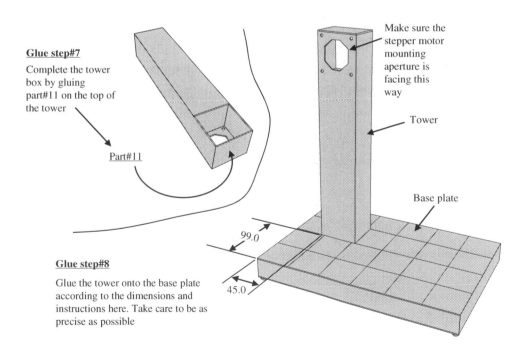

Glue step#7

Complete the tower box by gluing part#11 on the top of the tower

Part#11

Make sure the stepper motor mounting aperture is facing this way

Tower

Base plate

99.0

45.0

Glue step#8

Glue the tower onto the base plate according to the dimensions and instructions here. Take care to be as precise as possible

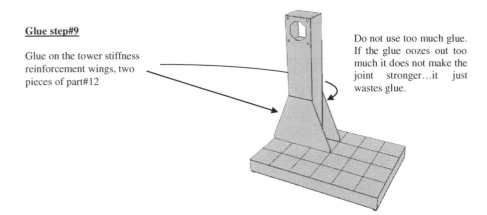

Glue step#9

Glue on the tower stiffness
reinforcement wings, two
pieces of part#12

Do not use too much glue.
If the glue oozes out too
much it does not make the
joint stronger…it just
wastes glue.

Okay that is the tower built and glued to the base plate. You should examine
the structure so far and realize that it is lightweight stiff and strong...why?
Well it is because shape, not material is the primary source of structural light-
ness, stiffness, and strength. It is not material that is the primary ingredient of
these characteristics. In other words, ingenuity is the key to engineering.

The next step is to bolt the stepper motor to the tower. You may have thought
that how can "mere" flimsy cardboard support a heavy motor. When we
screw the motor onto the tower you will see that cardboard *can*, in fact, provide a very effective mounting support. Now, before you fix the motor to the
tower, you need to use a flat needle file to file a "flat" onto the stepper motor
shaft. The flat is important because you should never tighten a screw onto a
shaft in order to secure a flange from turning. A flange is a mounting connector for the hitting arm to be built next. The flat ensures that (i) the outside
cylinder remains smooth and (ii) will effectively transfer torque to the shaft.
When you file the flat make sure you do not let iron filings enter the nearby
bearing and make sure the shaft does not rotate. It is a tricky procedure, you
will need patience.

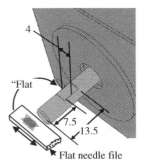

Flat needle file

To fix the motor, you will need four M4 stainless steel screws, crosshead or hex head (definitely not countersink), length 14 mm (or 16 mm) plus four stainless steel diameter 4 mm washers and four stainless steel M4 locknuts. Stainless steel is used because it looks good and doesn't rust and locknuts are used because a stepper motor will shake loose nuts if they are not locked. Feed the screws in from the back of the tower. It is tricky to do; you will need patience. Then feed the washer onto the screw from the front, see the following diagrams and then screw on the locknut. You will need to hold the screw head from turning with a screwdriver or hex key while you tighten the locknut. Do not tighten too much because you will damage the cardboard. Also tighten the screws progressively. Do not tighten just one screw while the other three are loose.

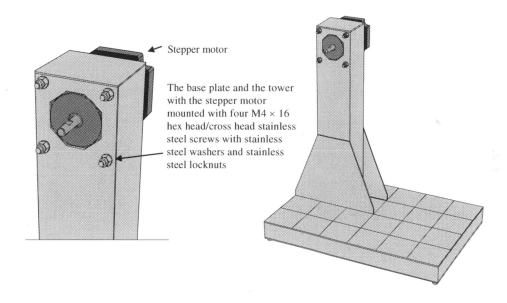

Stepper motor

The base plate and the tower with the stepper motor mounted with four M4 × 16 hex head/cross head stainless steel screws with stainless steel washers and stainless steel locknuts

The next stage is to build the hitting arm.

6.3 **MAKING THE HITTING ARM**

The hitting arm acts like a golf club that hits the ping-pong ball. Here are its drawing plans.

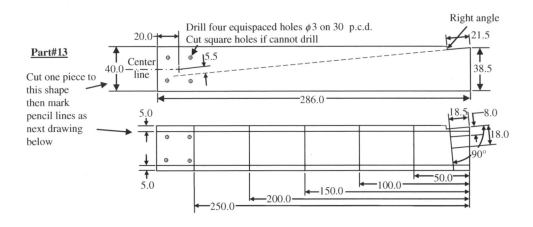

Part#13

Cut one piece to this shape then mark pencil lines as next drawing below →

Drill four equispaced holes ⌀3 on 30 p.c.d.
Cut square holes if cannot drill

Right angle

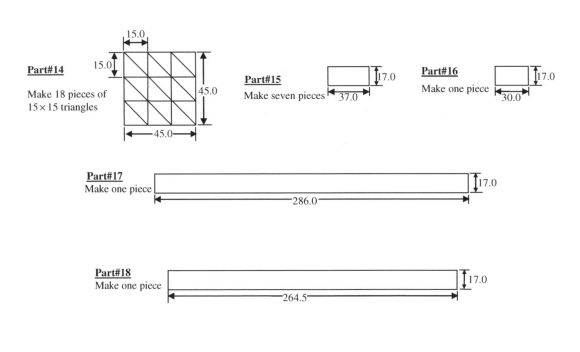

Part#14

Make 18 pieces of 15×15 triangles

Part#15

Make seven pieces

Part#16

Make one piece

Part#17
Make one piece

Part#18
Make one piece

Part#19
Make one piece
This part is based on part#13 so use its dimensions from drawing above

Same dimensions as part#13 above

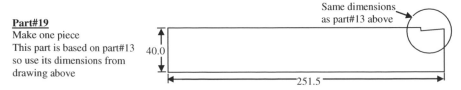

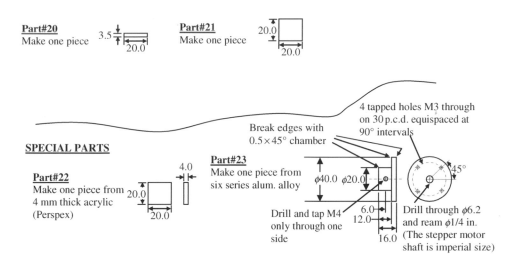

Part#20
Make one piece
3.5
20.0

Part#21
Make one piece
20.0
20.0

SPECIAL PARTS

Part#22
Make one piece from
4 mm thick acrylic
(Perspex)
20.0
4.0

Part#23
Make one piece from
six series alum. alloy

$\phi40.0$ $\phi20.0$

Break edges with
$0.5 \times 45°$ chamber

4 tapped holes M3 through
on 30 p.c.d. equispaced at
90° intervals

45°

Drill and tap M4
only through one
side

6.0
12.0
16.0

Drill through $\phi6.2$
and ream $\phi1/4$ in.
(The stepper motor
shaft is imperial size)

Now start gluing up the hitting arm.

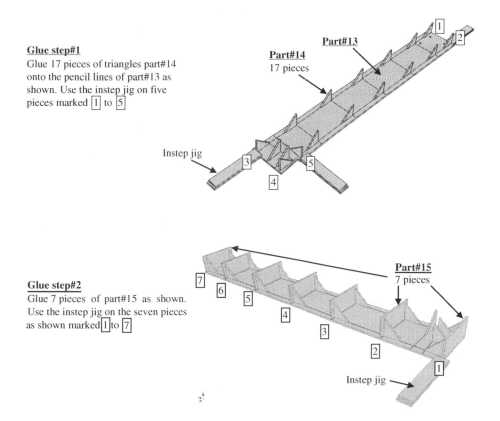

Glue step#1
Glue 17 pieces of triangles part#14
onto the pencil lines of part#13 as
shown. Use the instep jig on five
pieces marked 1 to 5

Part#13

Part#14
17 pieces

Instep jig

1
2
3
4
5

Glue step#2
Glue 7 pieces of part#15 as shown.
Use the instep jig on the seven pieces
as shown marked 1 to 7

Part#15
7 pieces

7
6
5
4
3
2
1

Instep jig

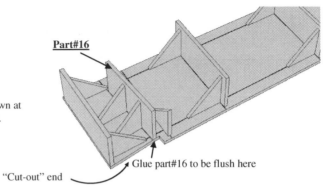

Part#16

Glue step#3

Glue one piece of part#16 as shown at the "cut-out" end of the structure.

Glue part#16 to be flush here

"Cut-out" end

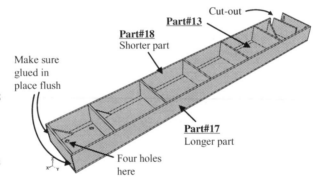

Cut-out

Part#13

Part#18
Shorter part

Make sure glued in place flush

Glue step#4

Glue two pieces, part#17 and part#18 on either side. Part#17 is longer than part#18. Make sure you glue the shorter one on the cut-out side of part#13. Make sure you glue the parts flush on the end with four holes

Part#17
Longer part

Four holes here

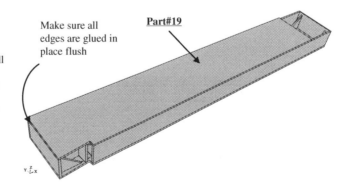

Make sure all edges are glued in place flush

Part#19

Glue step#5

Glue part#19 to close the structure. Make sure that all edges are glued in place flush. Note once again that 15 min after gluing that the structure becomes stiff and strong in bending and torsion.

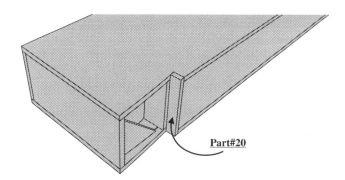

Glue step#6
Glue in place the very
small part#20

Part#20

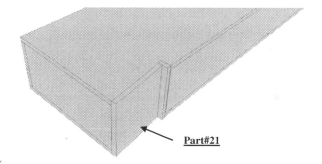

Glue step#7
Glue part#21 in place thus
closing the cut-out part

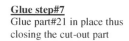

Part#21

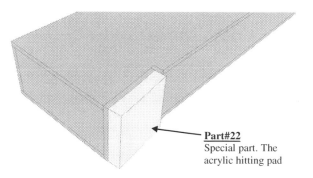

Glue step#8
Use double-sided stick tape to
affix special part#22 to the
cardboard part#19

Part#22
Special part. The
acrylic hitting pad

Special Part#22
It is a connecting flange for
the stepper motor shaft

Now attach Special Part#22 using
four M3 stainless steel screws,
crosshead, 6 mm long. Use a $\phi 3$
stainless steel washer between
each screw head and the
cardboard of the structure.
Tighten the screws progressively
into the M3 tapped holes of the
connecting flange

Make sure the M4
tapped hole is facing
in this direction to
make it easy to insert
and tighten the screw

Four pieces M3
stainless steel
screws,
crosshead,
6 mm long
with stainless
steel washers

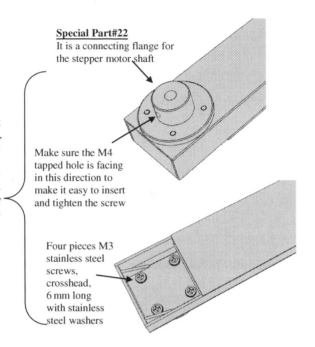

Now mount the arm onto
the stepper motor shaft

This is an M4 × 10 long crosshead stainless
steel screw. It is important that you tighten
the screw against the flat. To do this, feed
the flange onto the shaft without the screw
fitted and see by eye the flat when viewed
into the M4 tapped hole of the flange.
When you have done that, carefully keep
that position and screw in the M4 screw
until it tightens against the flat then release
the screw a little. You will be able to slide
the flange in and out a little as the screw
hits against the flat boundaries. When this
happens then you know for sure that the
screw bears against the flat so you can now
tighten the M4 screw. Beware that, later,
when you run the stepper motor the screw
may come loose and there will be a lot of
noise and the robot performance will be
greatly degraded.

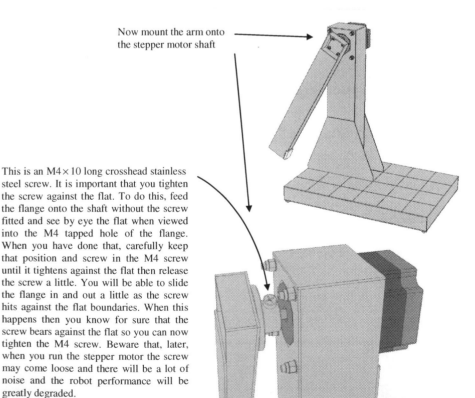

Okay that is the hitting arm finished and connected to the stepper motor. The next step is to build the "tee" on which is perched the ping-pong ball ready to be hit by the hitting arm.

6.4 **MAKING THE TEE**

Here are the drawings.

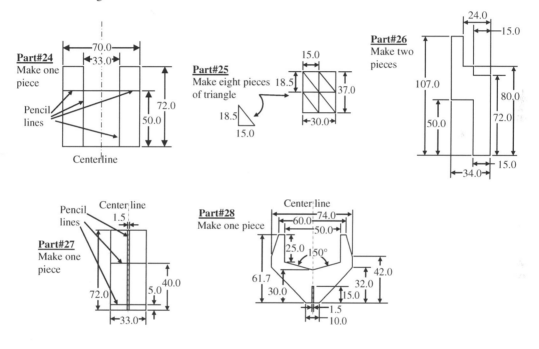

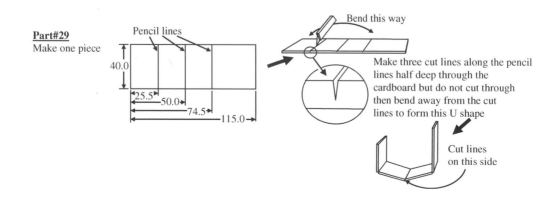

Make three cut lines along the pencil lines half deep through the cardboard but do not cut through then bend away from the cut lines to form this U shape

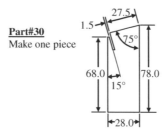

Part#30
Make one piece

Okay, that is the tee drawings done so now let us move on to its gluing and assembly.

Glue step#1

Glue four pieces of part#25 to part#24. Take care to:

(i)   Glue down the longer, not the shorter, side of the triangle.

(ii) Glue the triangles on the inside of the pencil lines.

(iii) Keep the triangles upright; using a T-square is a good technique.

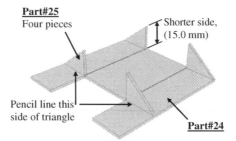

Part#25
Four pieces

Shorter side, (15.0 mm)

Pencil line this side of triangle

Part#24

Glue step#2

Glue two pieces of part#26 to the assembly. Make sure the parts#26 are flush at the bottom of the assembly. Trim your parts if necessary

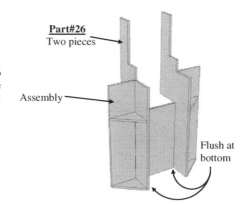

Part#26
Two pieces

Assembly

Flush at bottom

Glue step#3

Glue part#27 to the assembly and make sure the pencil marks are as shown. Make sure part#27 is glued when mating with parts##26 flush all round

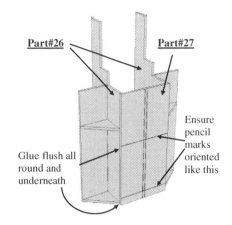

Part#26   Part#27

Ensure pencil marks oriented like this

Glue flush all round and underneath

Glue step#4

Glue 4 pieces of triangle part#25 onto part#27. This time glue the shorter of the two sides of the triangle. Glue the triangles such that a scrap piece of cardboard can fit snugly in the gap made by the triangles. Check after gluing that the assembly stands upright on a flat bottom.

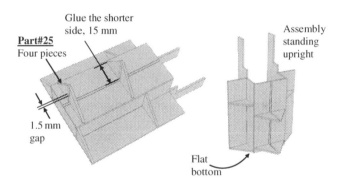

Glue the shorter side, 15 mm

Part#25
Four pieces

1.5 mm gap

Assembly standing upright

Flat bottom

Glue step#5

Put aside temporarily the assembly of glued step 4. Now glue the U-shaped part#29 to the mating part#28. Make sure the orientation is correct with the high side of part#29 on the right and the two parts mating flush at the front. The twoparts will be a snug fit and may be a little tight so you will have to coax the parts together and hold them a few minutes while the glue is drying There should be no gaps along the mating surfaces

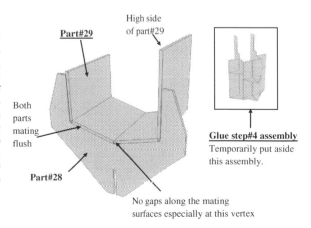

High side of part#29

Part#29

Both parts mating flush

Part#28

Glue step#4 assembly
Temporarily put aside this assembly.

No gaps along the mating surfaces especially at this vertex

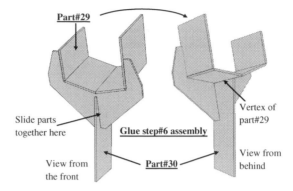

Glue step#6

Apply glue and slide part#30 onto part#28. Make sure that part#30 matches accurately with vertex of part#29 and these parts are also glued along the vertex. You will need to hold the parts together for a short time. This assembly is glue **step#6 assembly.**

Part#29

Slide parts together here

Glue step#6 assembly

Vertex of part#29

View from the front

Part#30

View from behind

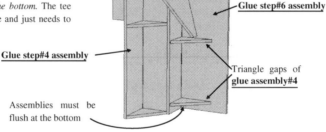

Glue step#7

Glue together glue step#4 assembly to glue step#6 assembly by gluing and sliding the two assemblies together into the triangle gaps of glue assembly#4. *Make sure the two assemblies are flush at the bottom.* The tee assembly is now complete and just needs to be glued to the base plate

Glue step#6 assembly

Glue step#4 assembly

Triangle gaps of **glue assembly#4**

Assemblies must be flush at the bottom

Glue step#8

The tee assembly should now be glued into position on the top surface of the base plate according to the diagram here

It is important that the hitting arm hitting pad strikes the ball with the following two specifications:

   (i) The launching angle is 40° (see Chapter 3).
   (ii) The ball collision point is in the center of area of the hitting pad.

The Hitting robot has been designed to fulfill these specifications due to the geometry of the hitting arm and the location of the tee that is glued to the base plate.

The diagrams on this page show geometrical and kinematic aspects of the hitting arm, the tee, and the ball.

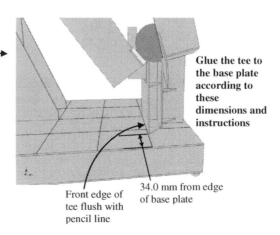

Glue the tee to the base plate according to these dimensions and instructions

Front edge of tee flush with pencil line

34.0 mm from edge of base plate

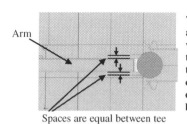

Arm

Spaces are equal between tee

Top view of arm at collision point with ball showing the arm hitting the ball in its center, i.e., arm is equally spaced between the tee

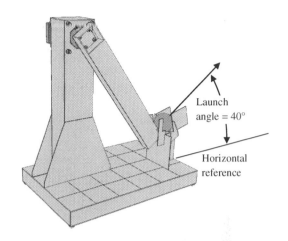

Launch angle = 40°

Horizontal reference

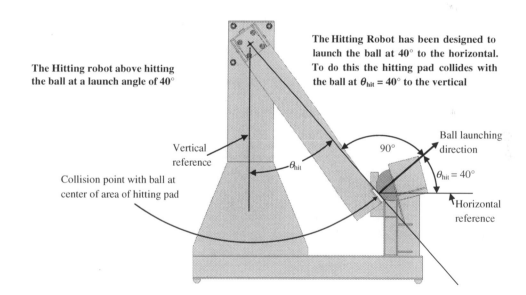

The Hitting robot above hitting the ball at a launch angle of 40°

The Hitting Robot has been designed to launch the ball at 40° to the horizontal. To do this the hitting pad collides with the ball at $\theta_{hit} = 40°$ to the vertical

Vertical reference

Collision point with ball at center of area of hitting pad

θ_{hit}

90°

Ball launching direction

$\theta_{hit} = 40°$

Horizontal reference

6.5 **MOUNTING THE ELECTRONICS**

That is the completion of the major components of the Hitting robot. We now need to mount the electronics, microcomputer, and sensor then write the software code that brings the robot to life. There are three ancillary structures now to be built that mount (i) the H-bridge stepper motor electrical power driver, (ii) the microcomputer, and (iii) the sensor that datums the angle of the hitting arm.

The H-bridge used by the Hitting robot is the Keyes L298 bipolar driver that is described in Chapter 7 complete with its electrical wiring and sample code.

Here are the drawings of the ancillary structure to mount the Keyes system complete with a fan to cool its power transistors.

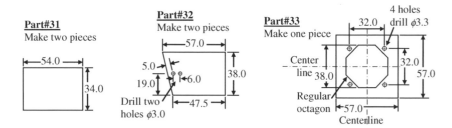

Part#31
Make two pieces

|←──54.0──→|

34.0

Part#32
Make two pieces

|←────57.0────→|

5.0→

19.0 |←6.0

38.0

Drill two |←──47.5──→|
holes ⌀3.0

Part#33
Make one piece

Center
line 38.0

Regular
octagon |←57.0→|
Centerline

4 holes
32.0   drill ⌀3.3

32.0
57.0

Just four gluing steps are required to create a mount for the H-bridge driver cooling fan. The parts are glued to structure that has been built so far as follows.

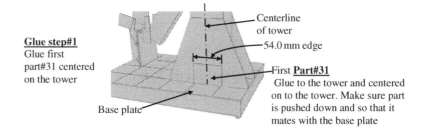

Glue step#1
Glue first
part#31 centered
on the tower

Centerline
of tower

54.0 mm edge

First **Part#31**
Glue to the tower and centered on to the tower. Make sure part is pushed down and so that it mates with the base plate

Base plate

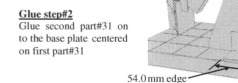

Glue step#2
Glue second part#31 on to the base plate centered on first part#31

Second part **Part#31**
Glue to the base plate and centered on the first **part#31**. Make sure it is pushed in to mate with the first part

54.0 mm edge

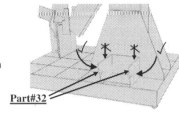

Glue step#3
Glue two parts of
part#32 as shown

Part#32

Two parts of **Part#32**
Glue the two parts as shown. Make
sure you glue from the left and right
sides and not from above as shown
with arrows

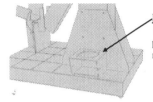

Glue step#4
Glue **part#33** on top of the
previously glued parts as shown

Part#33
Glue the part on top of the
previously glued parts and
mating with the tower

Now carry out drilling operation to enable mounting of the Keyes L298
driver circuit board.

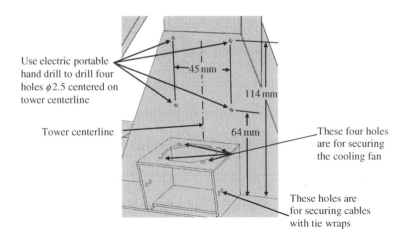

Use electric portable
hand drill to drill four
holes ϕ2.5 centered on
tower centerline

45 mm

114 mm

Tower centerline

64 mm

These four holes
are for securing
the cooling fan

These holes are
for securing cables
with tie wraps

Next step is to make the microcomputer mounting for the Basic Stamp
Board of Education. Here are the drawings.

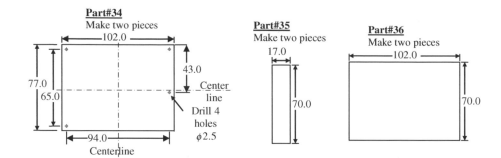

Part#34
Make two pieces
102.0
77.0
65.0
43.0
Center line
Drill 4 holes
φ2.5
94.0
Centerline

Part#35
Make two pieces
17.0
70.0

Part#36
Make two pieces
102.0
70.0

Now construct the microcomputer mounting assembly like this.

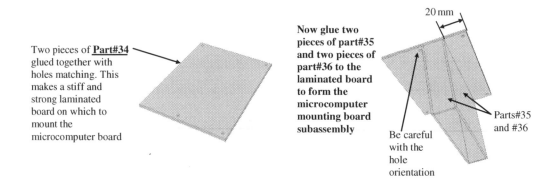

Two pieces of **Part#34** glued together with holes matching. This makes a stiff and strong laminated board on which to mount the microcomputer board

Now glue two pieces of part#35 and two pieces of part#36 to the laminated board to form the microcomputer mounting board subassembly

20 mm

Be careful with the hole orientation

Parts#35 and #36

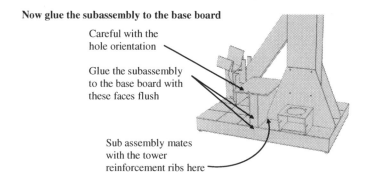

Now glue the subassembly to the base board

Careful with the hole orientation

Glue the subassembly to the base board with these faces flush

Sub assembly mates with the tower reinforcement ribs here

Now screw the cooling fan, the Keyes board, and the Basic Stamp board to the structure. Screw straight into the cardboard holes by treating the screws as "self tapping" screws. You need a firm but gentle force to get the thread started. Then when you feel the screw has started to cut into the cardboard, ease off on the force, and concentrate on feeling the torque applied to the screwdriver. As the parts are pulled together by the screw, the torque will increase. Be careful at this point and when the torque increase starts to decrease then stop tightening the screw. If you keep tightening, the cardboard self-tapped thread will strip and you will need to do a repair job which is unwanted extra work. The following photograph shows the three components screwed to the structure.

Parallax Basic Stamp Board of Education with USB programming port fitted with Basic Stamp 2 module but good idea to uprate to Basic Stamp2px or BS2sx module. Board is powered by 9 V rechargeable PP3 battery Board is screwed to mounting assembly by four brass standoff pillars same as used for the Keyes H-bridge board.

Keyes L298 H-bridge driver board screwed to tower using brass standoff pillars with M3 screws that are supplied with the board

Red and black wires of the cooling fan. These wires poach 12 V power supply from the Keyes board

NMB 1606KL-04W-B30 12 V DC 0.09A Brushless motor cooling fan held in place with four stainless steel screws M4 × 20 mm long, cross head. Self tap into the ϕ3.3 holes. Make sure the polarity of the fan is to blow air upward onto the circuit board above.

Make sure you connect the Basic Stamp board ground, V_{ss} to the ground of the Keyes board, GND. It is a common mistake by students not to connect grounds together of interconnected circuits. Remember you are dealing with floating voltage sources, for example, the battery supply to the basic Stamp board in the photo so that is why the grounds should be interconnected. If this is not done you will wonder why your programming does not work.

See Chapter 7 for the circuit diagram of the Keyes L298 H-bridge driver. The driver is supplied with 12 V power and is controlled by the Basic Stamp microcomputer. Chapter 7 also shows these control connections. As regards the NMB fan, it is a fan that is driven by a brushless motor but the clever internal circuitry means that all you do is to connect to a 12 V supply. Since the Keyes driver is supplied with 12 V then you simply poach from its supply. You can see this in the photograph above. Look for the red and black wires that go up to the top of the Keyes board.

Ok now for the sensor. We use a QTI sensor and students should Google this device to find out how it works and why it is so-named. It is an optical reflective sensor that will be used to detect the location of the arm so it can be datumed which means that the stepper motor is given a starting angle of zero step count at a known position. We need to mount the sensor and to make a reflecting plate like this.

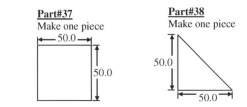

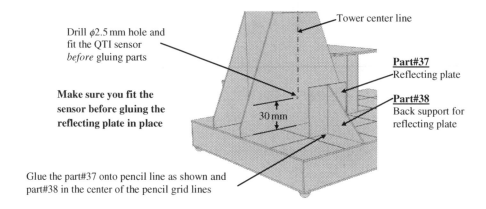

Part#37
Make one piece
50.0
50.0

Part#38
Make one piece
50.0
50.0

Tower center line

Drill ⌀2.5 mm hole and fit the QTI sensor *before* gluing parts

Make sure you fit the sensor before gluing the reflecting plate in place

30 mm

Part#37
Reflecting plate

Part#38
Back support for reflecting plate

Glue the part#37 onto pencil line as shown and part#38 in the center of the pencil grid lines

Part#37 is a reflecting plate for the QTI sensor that will be mounted in the drilled hole. It is necessary so that the sensor is not "staring into space" when the arm is not blocking the sensor's transmitting light path. The photo, left below, shows the system on the real robot. To fix the QTI sensor you need to drill a plastic, 10 mm long stand-off pillar down its center with a diameter 3 mm drill to remove its threads. Then secure the sensor and pillar with an M3 stainless steel screw, 16 mm long with a cross head that screws into two layers of cardboard of the tower. The inside end surface of the arm now requires a piece of white paper as a reflecting surface for the QTI sensor. The followings photos explain more.

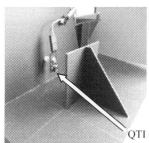

The photo, left, shows The QTI light sensor facing the reflecting plate. The photo, right, shows the sensor reflecting off a white piece of paper stuck to the end of the arm. As the arm sweeps past the QTI sensor, its microcomputer sense value drops to a low value indicating close high light reflectivity. When the arm moves away, then the sense value rises to a high value indicating low light reflectance.

QTI sensor mounted on a 10 mm plastic stand-off pillar

Glue a piece of white paper here at the end of the arm opposite to the sensor

We are very close now to programming the robot. Just before we do that, the QTI sensor needs to be wired up to the Basic Stamp. The following circuit diagram shows these connections and reminds us of the connections to the Keyes H-bridge. The circuit diagram serves as a programmer's pin connection diagram which is all a programmer needs to know for writing code. A programmer does not need to know about power connections and electrical circuits; he/she only needs to know the pin connections to the microcomputer.

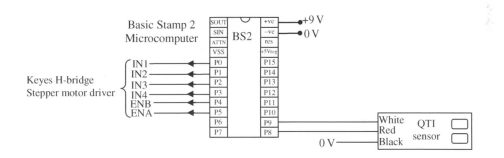

6.6 **PROGRAMMING THE HITTING ROBOT**

So, on completion of the sensor wiring we are ready for programming. Mechatronics design is largely concerned with "delayed gratification" since a lot of work and patience goes in to making the M (Mechanical) and E (Electrical, Electronics) components of the MEChatronics discipline. The C (Control and Computing) component has to wait until the ME hardware has been built and tested. The "tested" part is not to be underestimated since there are always things that do not work or do not go to plan or have to be redesigned or rebuilt. Anyway, when the time is right for the real-time programming of your robot it's a great feeling, like that of becoming an Emperor/Empress, after all the hard work creating hardware, who can sit back and command the robot to carry out his/her commands all under finger tip programming control that brings the robot to "life."

To complete our design of the Hitting robot, a sample Basic Stamp 2 programme is listed below with comments that explain the programme and act as a tutorial for the PBasic programming language. Students should refer to the Basic Stamp help manual that is at the top right of the editor screen. Note that the L298 chip will cut out if it gets too hot so the code below switches off the current shortly after the stepper motor has finished its task.

```
'{$STAMP BS2}        'inform the editor that you are using the BS2 module
'This is the hitting programme
'list of control pins, see circuit diagram above
'p0 = IN1 (L298)
'p1 = IN2 (L298)
'p2 = IN3 (L298)
'p3 = IN4 (L298)
'p4 = ENB (L298)
'p5 = ENA (L298)
'p6 is a delay signal. The pulsout instruction is used to output a 2µs-resolution pulse to give a delay time
'p7 is not used
'p9 and p10 used for the QTI sensor
DIRL=%11111111      'set pins p0, p1, p2, p3, p4, p5, p6, p7 to out direction because this port is
                    dedicated 'to controlling the L298 Keyes driver and to create a time delay
a       VAR Word
i       VAR Word
j       VAR Word
k       VAR Nib
sense   VAR Word    'QTI sensor variable that detects the arm position by measuring light intensity
b       CON 21      'acceleration constant used in "hit" subroutine. Bigger number equals bigger
                    'acceleration
c       CON 1
d       CON 7 '5 is fastest FOR vertical position 8 is ok for 45deg

OUTL=%00100001      'ensure motor current is off for safety, ENA and ENB are set high to turn off L298
PAUSE 1000          ' wait 1 second before running the programme. This is to allow updated
                    programmes 'to be downloaded before the old one starts running
'START OF MAIN PROGRAMME
        a=1400              'This is the delay value to give a slow speed suitable for datuming
skip1: OSUB backdirn     'Go backwards to datum the arm position with the arm at the QTI sensor
```

```
          GOSUB readsensor        'got to the subroutine that checks to read the light intensity
          IF sense>1000 THEN skip1      'arm not in view so keep going backwards. Note higher number
                                        'means low light; lower number means high light and arm is
                                        in view
          PAUSE 100               'sense<1001 so arm is in view at datum point
                         wait for 0.1 second for arm to stabilise
          FOR i=1 TO 41           'drive arm backwards by 295° to above the tee without interfering
                                   with ball
          GOSUB backdirn          'backwards by 8 steps (will do this 41 times)
          GOSUB hit               'now accelerate forwards to hit the ball at maximum speed
          PAUSE 100               'wait for 0.1 s for arm to stabilise
          OUTL=%00100001          'job done so ensure motor current is off for safety
'END OF MAIN PROGRAMME
again1:   GOTO again1             'infinite loop to end the programme

readsensor:   HIGH 10            'read the reflected light intensity of the QTI sensor
              HIGH 9
              PAUSE 1
              RCTIME 9,1,sense    'resistance of the junction is an inverse measure of light
                                    intensity
              LOW 10
'DEBUG DEC sense,CR  'use debug to check values of sense. Make sure it is deactivated as a 'comment
                     when 'running programme in real time because debug take 5 ms of valuable
                     execution time
              RETURN   'return from subroutine to main programme

Hit:   'acceleration phase to maximum speed in 0.9 of a revolution (324°) to hit ball
       a=1000
       FOR i=1 TO 45   '45 loops of 8 steps/loop hence giving 360 x ½ steps, i.e. 324° of a revolution
       a=a-b           'variable, "a" is the periodic time between steps. Variable "b" is the
                        acceleration 'constant. Here we accelerate from below pull-in speed, a=1000 - b
                        to maximum speed,
                        'a = 1000-(45xb)
       OUTL=%00100011          ' ½ step phase 1
       PULSOUT 6,a             'delay (a x 2 μs) which is the period of the frequency of each step
       OUTL=%00101011          ' ½ step phase 2
       PULSOUT 6,a
       OUTL=%00101001          ' ½ step phase 3
       PULSOUT 6,a
       OUTL=%00101101          ' ½ step phase 4
       PULSOUT 6,a
       OUTL=%00100101          ' ½ step phase 5
       PULSOUT 6,a
       OUTL=%00110101          ' ½ step phase 6
       PULSOUT 6,a
       OUTL=%00110001          ' ½ step phase 7
```

```
PULSOUT 6,a
OUTL=%00110011          ' ½ step phase 8
PULSOUT 6,a
NEXT
'deceleration phase to slow back down to below pull-in speed, i.e. reverse the acceleration
procedure
FOR i=1 TO 45
a=a+b
OUTL=%00100011          ' ½ step phase 1
PULSOUT 6,a
OUTL=%00101011          ' ½ step phase 2
PULSOUT 6,a
OUTL=%00101001          ' ½ step phase 3
PULSOUT 6,a
OUTL=%00101101          ' ½ step phase 4
PULSOUT 6,a
OUTL=%00100101          ' ½ step phase 5
PULSOUT 6,a
OUTL=%00110101          ' ½ step phase 6
PULSOUT 6,a
OUTL=%00110001          ' ½ step phase 7
PULSOUT 6,a
OUTL=%00110011          ' ½ step phase 8
PULSOUT 6,a
NEXT
RETURN

backdirn:a=1000   'set a new periodic step delay, slightly faster but still inside pull-in speed
OUTL=%00110011          ' ½ step phase 8 start from phase 8 and go backwards, 7, 6, 5 etc
PULSOUT 6,a
OUTL=%00110001          ' ½ step phase 7
PULSOUT 6,a
OUTL=%00110101          ' ½ step phase 6
PULSOUT 6,a
OUTL=%00100101          ' ½ step phase 5
PULSOUT 6,a
OUTL=%00101101          ' ½ step phase 4
PULSOUT 6,a
OUTL=%00101001          ' ½ step phase 3
PULSOUT 6,a
OUTL=%00101011          ' ½ step phase 2
PULSOUT 6,a
OUTL=%00100011    ' ½ step phase 1
PULSOUT 6,a
RETURN
```

6.7 **PROBLEMS**

1. If you have the use of a high-speed camera, try to find out what goes on at impact. For example, how does the ball deform during impact and what is the speed of the ball immediately before and after impact?
2. You can add some interesting features such as sound and lights using colored LED's. This is an exercise in augmented reality.
3. Check using an oscilloscope, which is assumed more accurate than the Basic Stamp clock, the angular velocity of the hitting arm.

Chapter 7

Theory V: The Angular Displacement Servomechanism, The "Servo"

LEARNING OUTCOMES

1. Basic understanding of a DC electric motor feedback closed-loop control system.
2. Experience in creating digital waveforms to give an angle displacement versus time of the servo shaft.

7.1 OVERVIEW OF THE SERVO

The "integrated servomechanism" or "servo," Figs. 7.1 and 7.2, is a device containing all necessary components integrated into a compact box with only three electrical wires for its control and power input. It produces rotation of the circular horn shown in Fig. 7.1. This type of servo was first developed some 50 years ago for the actuation of flying control surfaces of model radio controlled (r/c) aeroplanes. Sometimes it is referred to as an "r/c servo." Typically four servos are used inside the fuselage of a model aeroplane to actuate independently the ailerons, rudder, elevator, and engine throttle.

An operator, standing on the ground with a radio transmitter, manipulates a four degree-of-freedom joystick system that sends signals to a radio receiver in the model aeroplane. These signals cause the four servos to respond individually to the joystick commands. The r/c servo has become a standardized product in terms of size and operation. Numerous manufacturers produce the device, for example, Futaba, Hi-Tec. A bigger, heavier, more powerful, and low-cost range is produced by JX-Servo. Because the servo was designed for flying machines, it is lightweight, compact and responds very quickly to an input demand. Its mechanical power output of up to a few 10's of watts and its ease of control mean that it is suitable for actuating lightweight table-top sized robots. To be more precise, the servo is an *angular displacement servomechanism* which means it is a device that will convert

Creating Precision Robots. https://doi.org/10.1016/B978-0-12-815758-9.00007-9

■ **FIG. 7.1** The integrated angular displacement servomechanism or "servo."

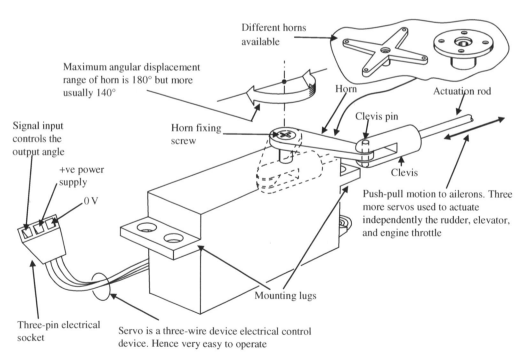

Different horns available

Maximum angular displacement range of horn is 180° but more usually 140°

Horn

Actuation rod

Clevis pin

Horn fixing screw

Signal input controls the output angle

+ve power supply

0 V

Clevis

Push-pull motion to ailerons. Three more servos used to actuate independently the rudder, elevator, and engine throttle

Mounting lugs

Three-pin electrical socket

Servo is a three-wire device electrical control device. Hence very easy to operate

■ **FIG. 7.2** Overview of the integrated servomechanism, otherwise known as a r/c servo. It is used as an actuator "muscle" for model aircraft, model cars, model boats, small robots, and mechanical systems.

a small power signal demand into an output shaft angular displacement, that is, position using feedback control.

The important thing to realize about a servo is that the output shaft will fight against any disturbance (within its torque limits) to maintain the output shaft angle at the demanded position (again within limits). It is not to be confused with a stepper motor. A servo and a stepper motor are not the same. The stepper motor will be described in this chapter.

The name, "servomechanism" is derived from "servo" = Greek for "slave" as in "servant" and "mechanism" = Greek for "machine." However a servomechanism can also describe a chemical or biological system, such as the temperature control of our bodies which regulates our internal body temperature at, $37 \pm 1°C$ ($98 \pm 2°F$). For example, if a human body carries out muscular exercise or if the weather is hot, its temperature starts to rise. Temperature sensors in the body respond by (i) diverting more blood to the skin surface which radiates more heat to the environment, (ii) the skin starts to sweat which gives a cooling effect due to the latent heat of evaporation, and (iii) breathing speeds up which, assuming the air temperature is low, cools the blood through the lungs. Conversely, if the body temperature drops then the muscles start to involuntary shake which adds heat to the body.

Another biological servomechanism concerns the use of an internal sugar sensor together with insulin which controls sugar levels in our bodies. Yet another servomechanism is an aeroplane autopilot. Autopilots are collections of servomechanisms mounted in the aeroplane, which keep the aeroplane on a desired heading, altitude, ground speed, and landing profile.

Servos are all around us. They pervade every part of our bodies and our lives. When we walk, our legs are placed in the correct positions through force impulses and similarly when we drink a cup of tea or write with a pen, the muscles of our arm and fingers are activated to produce coordinated movements. When we walk along a corridor we do not bump into the walls or other people because we provide differential effort to our legs to steer our bodies in the correct direction. When we drive a car, we need to maintain the car in the correct lane by providing differential motion of the steering wheel. The beauty of the servo is that only a small energy signal is required to control an output, for example, our brains use tiny amounts of electrical energy, manifested as a thought process, to control powerful muscles, such as our leg muscles to enable walking.

7.2 **BLOCK DIAGRAM OF THE INTEGRATED SERVOMECHANISM**

The key components of the integrated servomechanism can be summarized in the following block diagram, Fig. 7.3.

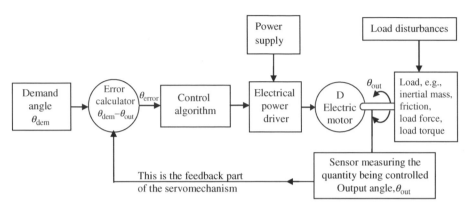

This book is not a book on control theory. There are many books on this subject so we will just summarize the functions of each block in the control system of Fig. 7.3.

The following briefly describes each block.

1. **Demand angle, θ_{dem}.** This is a very small power signal representing the value that we want, θ_{out}, to become as quickly as possible.
2. **Sensor measuring the quantity being controlled**. The integrated servo uses a potentiometer to measure the output angle, θ_{out}, you have to measure a quantity in order to control it.
3. **Error calculator, $\theta_{dem} - \theta_{out}$.** A good servomechanism will quickly reduce the error to zero or to a small value all the time. If the error is zero then $\theta_{dem} = \theta_{out}$, which is the central idea of a servomechanism.
4. **Control algorithm**. This block is the subject of hundreds of highly mathematical textbooks concerning control theory and has become very complicated. The block, nowadays, uses a digital computer. The error from the previous block, the error calculator, is amplified by a gain constant in order to fool the system into thinking the error is bigger than what it really is. This has the effect of producing a quick response in forcing, θ_{dem} to become equal to θ_{out} and also to provide a high resistance or stiffness against load disturbances. However, if the error is amplified too much then the response is too fast and, θ_{out}, starts to oscillate which requires a modification to the algorithm, usually a damping constant, to eliminate the oscillation. The error can be positive or negative depending on whether the output angle, θ_{out}, is less than or more than the demand angle, θ_{dem}, respectively. The sign of the error drives the electric motor in different directions, for example, CW or CCW, such that the error is reduced close to zero.

5. **Electrical power driver**. The electrical power driver can drive the electric motor with both a negative and a positive voltage with the necessary required electrical current. The bipolar nature of the driver voltage means that the electric motor can change its direction depending on the sign of θ_{error} such that the error is decreased to zero. The magnitude of the voltage from the power driver is proportional to the magnitude of the error voltage which means that the electric motor will run faster proportional to error value.

6. **DC electric motor**. The motor includes a reduction gearbox that provides a high torque output. A DC electric motor changes direction if the polarity of the input drives voltage changes. The speed of a motor is proportional to the input voltage under no-load conditions but if the load increases such that the motor is stalled, that is, its angular velocity is zero then the motor torque is proportional to the drive current.

7. **Sensor measuring the quantity being controlled, that is, the output angle, θ_{out}**. To control a system, it is necessary to measure the quantity being controlled to determine if it is equal to the demanded quantity.

7.3 **INSIDE THE INTEGRATED SERVO**

In order to explain how the integrated servo works, we will take it apart, part by part, and describe each part in more detail. Let us start with Fig. 7.4 which shows what is inside the servo.

Description of Each Component of the Integrated Servo

1. The quantity being controlled, θ_{out}

The quantity being controlled, Fig. 7.5, is the output shaft angle, θ_{out}, which is the angle protracted between an arbitrary datum fixed to the servo body and the clevis pin hole in the horn. The output shaft angle can be measured in a CW or CCW direction from a given datum. For example, in Fig. 7.5 the datum is set such that the shaft angle is measured in a CW direction. Alternatively, you can set the datum at a new position, Fig. 7.6, and measure the horn angle from this new datum. However, do not forget that you have to unscrew the horn and re-position it at a new angle such that the mid-range horn position from CW end stop to CCW end stop lies in the middle of the required range. How to do this will now be described.

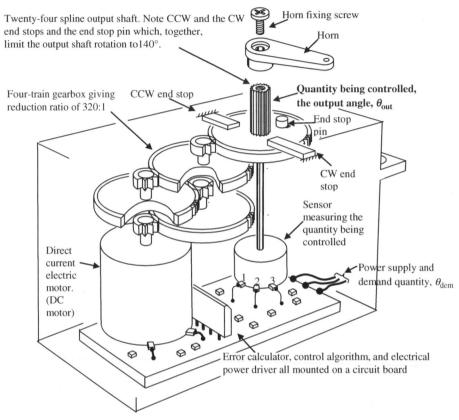

Twenty-four spline output shaft. Note CCW and the CW end stops and the end stop pin which, together, limit the output shaft rotation to140°.

Horn fixing screw

Horn

Four-train gearbox giving reduction ratio of 320:1

CCW end stop

Quantity being controlled, the output angle, θ_{out}

End stop pin

CW end stop

Sensor measuring the quantity being controlled

Direct current electric motor. (DC motor)

Power supply and demand quantity, θ_{dem}

Error calculator, control algorithm, and electrical power driver all mounted on a circuit board

■ **FIG. 7.4** Inside the integrated servo showing components responsible for controlling the output angle, θ_{out}.

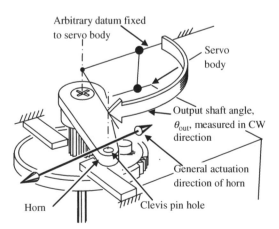

Arbitrary datum fixed to servo body

Servo body

Output shaft angle, θ_{out}, measured in CW direction

General actuation direction of horn

Horn

Clevis pin hole

■ **FIG. 7.5** The quantity being controlled, output shaft angle, θ_{out}.

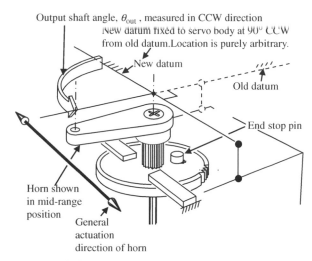

Output shaft angle, θ_{out}, measured in CCW direction
New datum fixed to servo body at 90° CCW
from old datum. Location is purely arbitrary.

New datum

Old datum

End stop pin

Horn shown
in mid-range
position

General
actuation
direction of horn

■ **FIG. 7.6** Redatuming the horn.

Datuming the Horn

A point to note about the horn is that it can be removed from the 24-spline output shaft and then replaced at a different angle from its previous position which is an integer multiple of 15 degrees (=360 degrees/24 splines). This means that you will be lucky for the mid-range horn position from CW end stop to CCW end stop will coincide with the middle of the required range. The worst case error will be 15 degrees/2, that is, 7.5 degrees. The principle is simple but difficult to grasp. To make the principle clear the steps are shown in Fig. 7.7.

2. The sensor measuring the quantity, θ_{out} being controlled

Fig. 7.8 shows a commonly used, but nevertheless ingenious, sensor called a rotary electrical potentiometer that is used to measure the quantity being controlled. This quantity is the angle of the output shaft, θ_{out}. A connecting rod faithfully transmits the shaft angle directly to the potentiometer.

The potentiometer contains a resistance track and a wiper which is electrically conducting and touches the resistance track with low force. The wiper is fixed to the connecting rod but electrically isolated from it. The linear resistance track has 0 and +5 V connected at its ends via two terminals, that is, terminal 1 and terminal 3, respectively. The voltage, v_{out}, picked off by the wiper is connected to a third terminal, that is, 2. The voltage, v_{out}, is an analogue of the output shaft angle, θ_{out}.

Fig. 7.9 shows an experimental set up devised to show how angle, θ_{out}, is transformed into a voltage. The potentiometer is unsoldered from the circuit

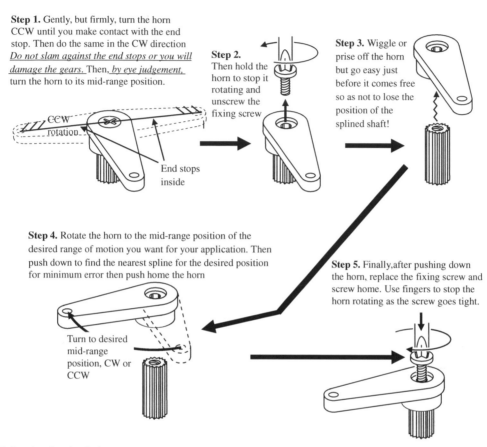

Step 1. Gently, but firmly, turn the horn CCW until you make contact with the end stop. Then do the same in the CW direction *Do not slam against the end stops or you will damage the gears.* Then, *by eye judgement,* turn the horn to its mid-range position.

ccw rotation

End stops inside

Step 2. Then hold the horn to stop it rotating and unscrew the fixing screw

Step 3. Wiggle or prise off the horn but go easy just before it comes free so as not to lose the position of the splined shaft!

Step 4. Rotate the horn to the mid-range position of the desired range of motion you want for your application. Then push down to find the nearest spline for the desired position for minimum error then push home the horn

Turn to desired mid-range position, CW or CCW

Step 5. Finally,after pushing down the horn, replace the fixing screw and screw home. Use fingers to stop the horn rotating as the screw goes tight.

■ **FIG. 7.7** Steps in redatuming the horn.

board and three wires are soldered onto terminals 1, 2, and 3 of the potentiometer. Next, the servo is fixed to the bench and a large diameter (150 mm) protractor is screwed to a circular horn in order to measure the output shaft angle. The output shaft angle, θ_{out}, is read by eye from a pointer fixed to the bench.

Fig. 7.10 shows the result of the experiment. The first feature to notice is that the relationship between output voltage, v_{out}, and shaft angle, θ_{out}, is linear:

$$v_{out} = 1.25 + \frac{2.5}{180} \times \theta_{out} \ (°) \tag{7.1}$$

Thus the control algorithm circuitry can calculate the output angle, θ_{out}, with an analogue to digital converter, ADC, using Eq. (7.2) to convert the digital number to the output angle.

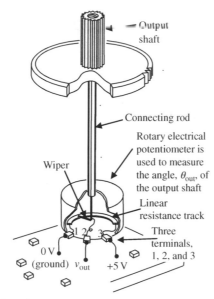

■ **FIG. 7.8** Rotary electrical potentiometer.

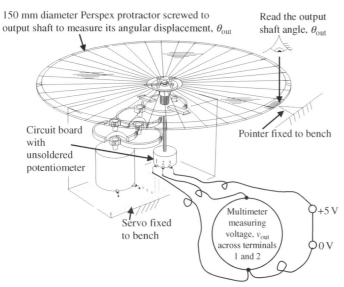

■ **FIG. 7.9** Experimental rig to illustrate method of transducing shaft angle, θ_{out} to a voltage that is proportional to angle.

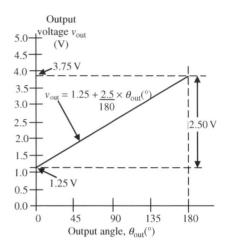

■ **FIG. 7.10** Relationship between potentiometer output voltage and output angle.

The servo is now in a position to compare the output angle, θ_{out}, with the demand angle, θ_{dem} to give θ_{error} as given by $\theta_{error} = \theta_{dem} - \theta_{out}$.

The error which has a sign, +ve or −ve, and magnitude is then used by the control algorithm to drive an electrical power driver, usually a chopper H-bridge power drive circuit (see this chapter). The electrical power driver will drive the electric motor CW or CCW depending on the sign of the error and will drive the motor faster or with more torque where both are proportional to the magnitude of the error. The motor is always driven in the correct direction in order to decrease the error.

Rearranging Eq. (7.1), to obtain θ_{out} as a function of v_{out}, the following is obtained:

$$\text{Output shaft angle,}\, \theta_{out}\,(^\circ) = -90 + \frac{180}{2.5} \times v_{out} \qquad (7.2)$$

The end result of the servo is to rotate the output shaft angle to the desired input demand angle, θ_{dem}. We now conclude the description of the components that make up the servomechanism. Attention now is turned to the demand signal, θ_{dem} and the signal control of the servo.

3. Demand quantity, θ_{dem}

The servo has a three-wire connection cable, Fig. 7.11, usually colored wires of black, red, and white. The black wire is the 0 V input, the red is the +4.8 to +7.2 V power input, and white is the TTL angle demand signal.

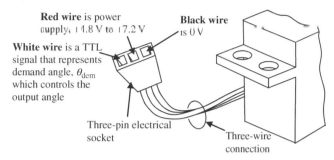

Red wire is power supply, +4.8 V to +7.2 V

Black wire is 0 V

White wire is a TTL signal that represents demand angle, θ_{dem} which controls the output angle

Three-pin electrical socket

Three-wire connection

■ **FIG. 7.11** Three-wire connection to servo.

7.4 MICROCOMPUTER SIGNAL CONTROL OF THE SERVO

The demand quantity, θ_{dem} is sent to the white wire. It is an electronic signal that represents the required angle of the output shaft. The question is: how do you represent an angle demand range from 0 to 90 degrees or 180 degrees? The answer is with a +5 V signal pulse width from 1 to 2 ms in duration. A pulse width is used because a high-low two-state signal can only communicate two values. For example, a low signal could indicate a demand angle of 0 degree, and a high signal could indicate 90 degrees. This is of little use because how would it be possible to indicate an angle of, for example, 35.6 degrees? One answer is to use the time that the signal is in the high state as representing the angle. This time in the high state is known as the, "pulse width" and such a signal is shown in Fig. 7.12. The signal is called a pulse width modulated, pwm, TTL (transistor-transistor-logic) signal. It is an industry standard communication protocol that uses a +5 V signal to indicate a logic "1" and a 0 V signal to indicate a logic "0." These two signals are also known as "high" and "low," respectively. The voltages that represent the two logic signals vary according to the latest integrated circuit technology. As a rough guide, the logic high voltage can drop as low as 3 V but should not exceed 5.25 V. The logic low voltage should not exceed 0.8 V and not be less than –0.25 V. (The latest, low power, battery compatible, technology uses 3.5 V for logic "1.")

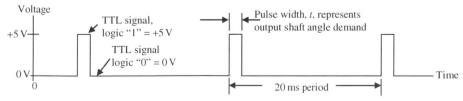

Voltage

+5 V

TTL signal, logic "1" = +5 V

TTL signal logic "0" = 0 V

Pulse width, t, represents output shaft angle demand

0 V

0

20 ms period

Time

■ **FIG. 7.12** Pulse width modulated TTL signal representing the demand quantity, θ_{dem}.

Machines that are controlled with digital microcomputers rely on communicating information at a certain rate, known as the "update rate," or refresh rate. For example, military fighter aircraft will communicate control data at a fast rate, for example, a 1 kHz update rate, that is, an update period of 1 ms. A control system for an automated greenhouse system that has to open and close windows according to weather conditions would have an update rate of, for example, 1 Hz, that is, an update period of 1 s. Another example is the picture refresh rate of standard television technology which is 50 Hz, that is, 25 ms update rate. The nominal update rate chosen for servo technology is also 50 Hz or 20 ms. In reality, the update rate for servos can vary between 10 and 25 ms, with a preferred value of 20 ms, Fig. 7.12. You need to remember that the pulse width encodes the demand angle, θ_{dem}, *not the period*. The period does not change the demand angle. *If you keep the pulse width constant and vary the period then the demand angle stays the same.* Furthermore reducing the period does not necessarily mean a faster response from the servo because it is the response of the electric motor that determines the upper limit response speed. Some servos will go down to an absolute minimum of 7 ms period but it is better to run safe at 10 ms. It is important to realize that if the period drops below the absolute minimum for a significant duration, for example, 2 min, or, much less for high performance servos, then permanent damage will be done to the servo due to the internal electronics locking the DC motor into current saturation. The same occurs if the pulse width demands an angle that is beyond either end stop; the electric motor will try to force the output shaft through the stops and as a result there will be high current draw and the servo electronics or motor will burn out. The latest servos do not suffer from this problem and if, either the period or the pulse width is out of range then the motor is switched off into a safe mode.

At the other extreme, if the period is greater than 25 ms then no damage will be done to the servo but it will begin to lose performance with a falloff in torque and oscillatory kicks will be felt at the output shaft on receipt of each pulse. As the period gets even longer then eventually the shaft torque becomes so feeble as to render the servo ineffective. The reason is due to the design of the servo electronics and is an historical safety feature that switches off the servo if the signal is lost.

Servo Calibration

The integrated servo is designed to accept angle demand pulse widths that vary from 0.5 to 2.5 ms with the relationship between angle demand and pulse width as shown in Fig. 7.13.

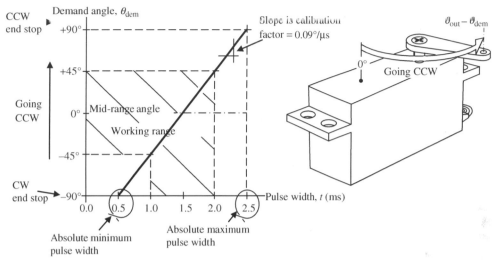

It is important to note that most servos, but not all, require a train of pulses to maintain an angle demanded output even if the angle demand is constant. If the angle demand is constant a train of constant pulse width pulses must be sent to the demand pin. For example, Fig. 7.14 shows a method of demanding a constant angle with just one pulse that will not work with the standard range of Futaba servos.

In fact, the above waveform will give an initial kick to the output shaft on receipt of the pulse then the servo will become inactive and limp. The integrated servo is designed to receive a train of pulses to become active. Not sending pulses has the advantage that the servo will become inactive which is useful for putting robots into a standby sleep state taking only a quiescent current of approximately 10 mA.

The following waveform, Fig. 7.15, will work better but only for a short time then make the servo inactive which results in the output shaft becoming limp.

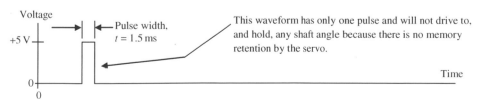

■ **FIG. 7.14** One pulse will not hold a demanded servo angle.

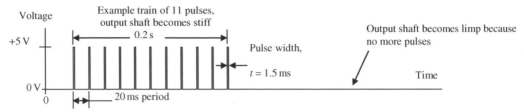

■ **FIG. 7.15** A train of pulses will hold a demanded angle but only for the duration of the train.

Becoming limp means that the output shaft will move if a torque is applied to it, otherwise it will stay where it is. To get to a demanded angle you must give enough pulses, which represents enough time, for the servo to react. Also the bigger the angular displacement then the longer time is required and thus proportionately more pulses. Remember also that the period does not alter the demanded angle; it is the pulse width that controls the demanded angle. Fig. 7.16 reinforces this point.

The reaction times to changes in demand displacements are the same for both waveforms. Programmers should not think that by decreasing the period that the performance will be increased. Instead, the speed of reaction of the output shaft is set fundamentally by the mechanical power output of the DC motor. It is also incorrect to think of the demand signal as being dependent on the duty cycle of the pwm waveform. For example, the above two waveforms have the same demand output shaft angle but the duty cycle

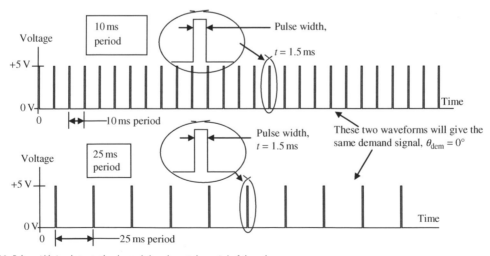

■ **FIG. 7.16** Pulse width is what sets the demanded angle; not the period of the pulse.

of the first waveform is, 1.5 ms/10 ms = 15%, and the second waveform is 1.5 ms/25 ms = 6%.

It is instructive to give two example waveforms and their microcomputer programming code that drives the output shaft through interesting angular profiles with respect to time. The first one is a square wave and the second one is a sawtooth waveform.

7.5 SQUARE-WAVE ANGLE DEMAND AND ITS PROGRAMMING CODE

Fig. 7.17 shows the demand angle requirement to give a 1 s periodic time square wave.

Fig. 7.18 shows a Basic Stamp2 microcomputer and servo circuit diagram that uses the Basic Stamp pin0 to generate the waveforms to control the servo to drive according to the square wave in Fig. 7.17.

Here is the Basic Stamp2 programming code to implement the 1 s periodic time square wave to drive the horn from −45 to +45 degrees and repeat. The BS2 has a clock period of 2 μs. Pin1 is used as a "dummy" output high resolution delay of 2 μs resolution.

```
i        VAR    word
again:   FOR i = 1 to 25              'generate 25 pulses with periodic time 20ms to give 0.5s pulse train
         PULSOUT 0,500               'one pulse x 1ms pulse width output from pin0 to drive to -45°
         PULSOUT 1,10000-500-450     'delay by 20ms-pulse width -administration time (to be calibrated)
         NEXT
         FOR i = 1 to 25              'generate 25 pulses with periodic time 20ms to give 0.5s pulse train
         PULSOUT 0,1000              'one pulse x 2ms pulse width output from pin0 to drive to +45°
         PULSOUT 1,10000-500-450     'delay by 20ms-pulse width -administration time (to be calibrated)
         NEXT
         GOTO again                  'repeat the cycle thus generating a square wave demand
```

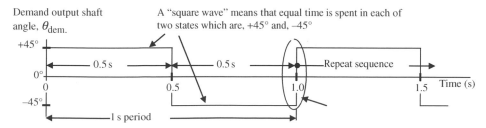

■ **FIG. 7.17** Square wave servo output shaft angle demand with respect to time.

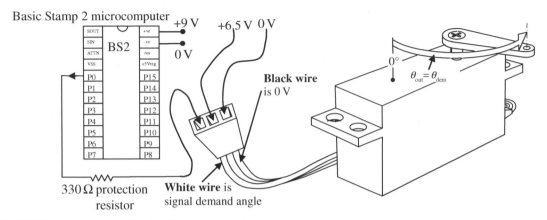

■ **FIG. 7.18** Wiring diagram between Basic Stamp2 microcomputer and servo.

You need to view the waveform on the oscilloscope as well as to watch the response of the servo in order to appreciate what is going on. Fig. 7.19 illustrates what you should see. You will need to play with the time base value. For example, the time base set to 200 ms, you will see a series of "spikes" that does not make the pulse width clear. However, with the time base set to 5 ms you will see two pulses and their pulse width can be seen clearly and with the time base set to 1 ms you can see clearly the pulse width changing every ½ s from 1 to 2 ms. Also the oscilloscope will help you calibrate the "administration" time which is the time to carry out the NEXT loop. The idea is that the administration time number, that is, 450 in the code above, is adjusted to make the periodic time equal to 20 ms.

You should now reduce the square wave period to see its effect. You will notice that the servo takes time to rotate from –45 to +45 degrees and vice

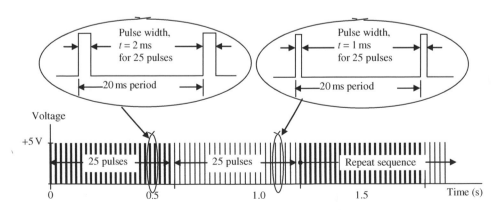

■ **FIG. 7.19** The train of pulses required to give a square wave output shaft angle.

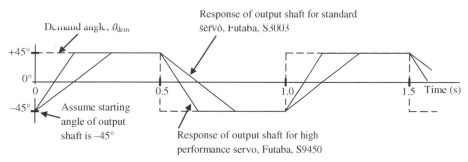

■ **FIG. 7.20** The output shaft angle response to the train of pulses for two types of servo.

versa. Also different servos will give different performance. For example, the low cost, low-performance S3003 Futaba servo will take longer time to travel through the demanded range than the high performance, more expensive, more responsive S9450 Futaba servo, Fig. 7.20.

7.6 SAWTOOTH WAVE ANGLE DEMAND AND ITS PROGRAMMING CODE

Fig. 7.21 shows a symmetrical sawtooth demand angle waveform that demands an angle range from –45 to +45 degrees with an equal positive and negative slope linear change in angle with respect to time. The periodic time of the sawtooth waveform is 1 s.

We have now to plan a strategy for programming such a practical waveform shape. First of all, 25 steps are needed to step from –45 to +45 degrees and 25 steps back down to –45 degrees, Fig. 7.22. This is represented in Basic Stamp code as a PULSOUT number ranging from 500 to 1000 and back down to 500, respectively. So each PULSOUT step is $(1000 - 500)/25 = 20$ which is the step size. Hence, the PULSOUT number should start at 500 and then increment to 520 then 540 and so on up to 1000 and then should decrement to 980 and so on back down to 500. Remember each PULSOUT instruction occurs every 20 ms, so 25 steps are required to

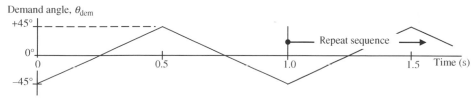

■ **FIG. 7.21** A symmetrical sawtooth demand angle waveform.

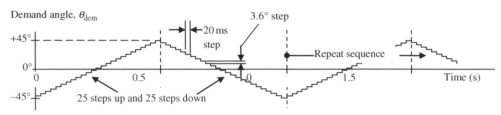

■ **FIG. 7.22** The practical implementation of the triangular demand output shaft angle.

change the angle demand from –45 to +45 degrees in 20 ms × 25 = ½ s. Also, 25 steps are required to drive the demand angle through 90 degrees range so each step represents 90 degrees/25 steps = 3.6 degrees/step. This lack of smoothness is a consequence of digital systems. It can be made smoother by changing to a 10 ms update rate where the step size would be 1.8 degrees but we will stick to the standard 20 ms rate.

In an alternative format, Fig. 7.23 shows the sawtooth waveform in terms of pulse width rather than demand angle. We are now ready to programme the Basic Stamp to undertake the sawtooth waveform, so here is the programme to that:

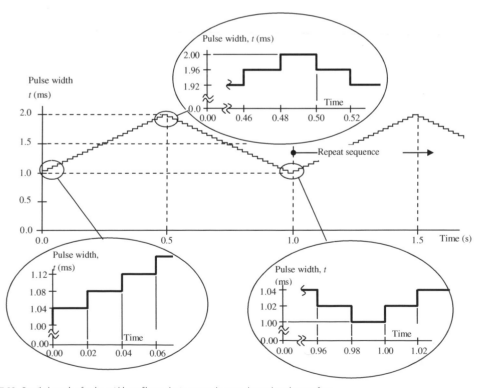

■ **FIG. 7.23** Detailed graph of pulse width profile to obtain sawtooth servo demand angle waveform.

```
pulsewidth       var   word
stepincrement    CON   20
again:           FOR pulsewidth=500 TO (1000-stepincrement)     STEP stepsize
                 PULSOUT 0,pulsewidth
                 PULSOUT 1,10000-pulsewidth-450
                 NEXT
                 FOR pulsewidth=1000 TO (500+stepincrement)     STEP stepsize
                 PULSOUT 0,pulsewidth
                 PULSOUT 1,10000-pulsewidth-450
                 NEXT
                 GOTO again
```

Note in the above programme that the first FOR-NEXT loop starts at 500 and ends at (1000 − stepincrement)=980. This is because we do not want to repeat the PULSOUT 0,1000 instruction. In other words once we arrive at 1000 then we start decrementing and likewise for the lower value of 500. Similarly, the second FOR-NEXT loop ends at 500+stepincrement so as not to repeat the PULSOUT 0,500 instruction.

Once again, students should see these waveforms on the oscilloscope where you will see the waveforms in Fig. 7.24.

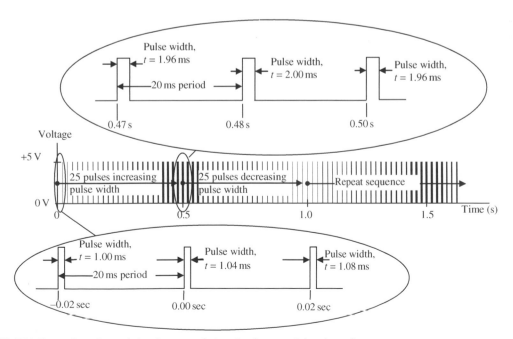

■ **FIG. 7.24** The actual waveform applied to the servo to obtain a triangular output shaft angle waveform.

You need to check that the periodic time of the pulses is 20 ms with the time base set to 5 ms. Then you should set the time base to 1 ms to see the pulse width expanding and contracting in width which is very revealing. At the same time watch and relate how the servo horn travels CW then CCW following the pulse widths. Many students are confused that the electric motor inside the servo reverses direction without being told to do so. The servo does the CW and CCW drive of the electric motor *automatically*; do you remember? It is built into the sign of the error which was discussed in Section 7.2 as the "error calculator." You need to appreciate that the pulse width demands the angle of the servo; it does not demand the direction of the motor; that is done automatically for you.

7.7 **PROBLEMS**

1. Write code that determines the Bode amplitude and phase response for the servo. This will mean that you need to work out how to create a sinusoidal drive waveform. You can accurately approximate such a waveform with a parabolic equation. The other way to do it is drive the servo with a square wave and to remember your Fourier theory which means that a square wave consists of the fundamental sinusoid plus harmonics. At high frequencies the servo acts as a low pass filter meaning that the servo cannot respond to high frequencies but can respond to the fundamental frequency.

2. Screw a long arm to the servo horn then write a programme to fix the servo at its mid-range position then use a spring balance to provide a torque to the horn. Now do an experiment to determine how the servo electrical current varies with CW applied torque. Now repeat the experiment in CCW direction. Now repeat the experiments yet again but this time measure the offset displacement as a function of applied torque.

Chapter 8

The 10-Ball Magazine Autoloader: Design and Construction

LEARNING OUTCOMES

1. Experience of increased level of construction complexity.
2. Understanding the integrated servo, its calibration and its programming.
3. Experience of increased level of programming complexity when the autoloader is fitted to the Hitter robot.

8.1 DESIGN DRAWINGS

Here are the design drawings of the cardboard parts to be made. We still use 1.5 mm thick cardboard.

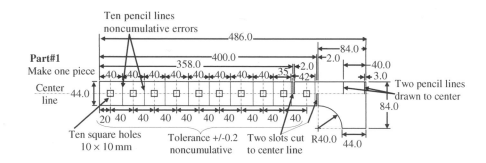

Part#2
Make one piece.
Copy the square hole dimensions from part#1.
This part is essentially a mirror image of part#1

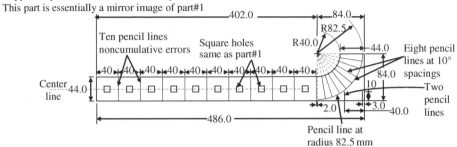

Ten pencil lines noncumulative errors

Square holes same as part#1

R82.5

R40.0

402.0

84.0

44.0

Eight pencil lines at 10° spacings

84.0

Center line   44.0

40 40 40 40 40 40 40 40 40 40

10

Two pencil lines

2.0   3.0   40.0

486.0

Pencil line at radius 82.5 mm

Part#3
Make one piece.
Copy the pencil line and square hole dimensions from part#2.

Note different width to part#1 and part#2

Nine pencil lines

Center line

41.0

400.0

Part#4
Make one piece.
Copy the pencil line and square hole dimensions from part#2.

Cut slot to the center line

Be Careful!
The slot is on this side

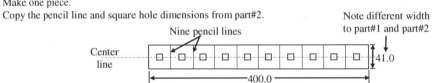

Once again be careful of this dimension

41.0

380.0

358.0

2.0

Center line

534.0

402.0

Side view of part#4
Bend the end to be as close to these dimensions as possible
so as to match the pencil line curve marked on part#2

R82.5

The bend is downwards

Part#5
Make 60 pieces of 10 × 10 triangle

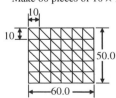

10

10

50.0

60.0

Part#6
Make one piece

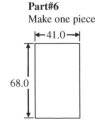

41.0

68.0

Part#7
Make four pieces

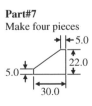

5.0

22.0

5.0

30.0

Part#8
Make four pieces

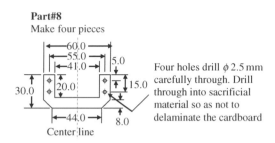

Four holes drill ϕ 2.5 mm carefully through. Drill through into sacrificial material so as not to delaminate the cardboard

Center line

Part#9
Make two pieces

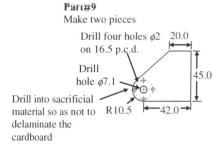

Drill four holes ϕ2 on 16.5 p.c.d.

Drill hole ϕ7.1

Drill into sacrificial material so as not to delaminate the cardboard

R10.5

Part#10
Make two pieces

Pencil line

Part#11
Make three pieces

Part#12
Make two pieces

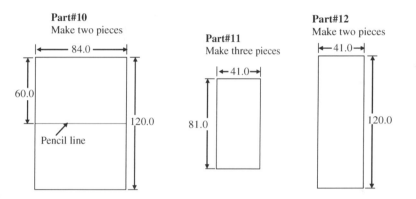

8.2 GLUING AND ASSEMBLY INSTRUCTIONS

Glue step#1

Glue 33 pieces of part#5 triangles to part#2. Once again use the instep jig

Use the instep jig to set the triangles inboard by 1.5 mm

Part#5 triangle

Glue the triangles on the radial lines and to match the curved line

Glue the triangles onto the pencil lines

Part#2

Do not forget these two triangles which are also set inboard with the instep jig and also glued on to the pencil lines

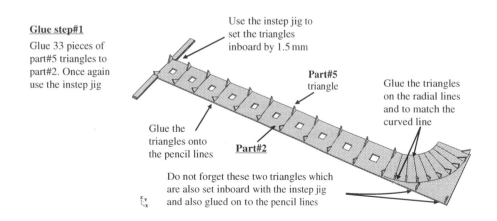

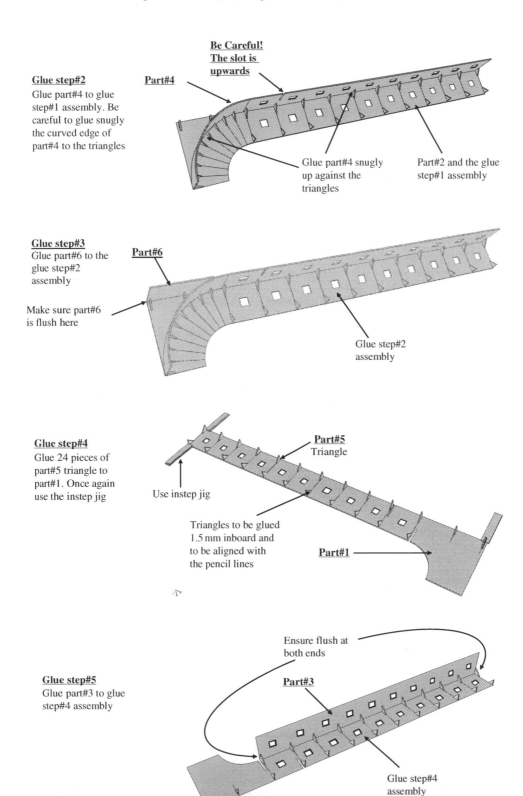

Glue step#2
Glue part#4 to glue
step#1 assembly. Be
careful to glue snugly
the curved edge of
part#4 to the triangles

Part#4

**Be Careful!
The slot is
upwards**

Glue part#4 snugly
up against the
triangles

Part#2 and the glue
step#1 assembly

Glue step#3
Glue part#6 to the
glue step#2
assembly

Part#6

Make sure part#6
is flush here

Glue step#2
assembly

Glue step#4
Glue 24 pieces of
part#5 triangle to
part#1. Once again
use the instep jig

Use instep jig

Part#5
Triangle

Triangles to be glued
1.5 mm inboard and
to be aligned with
the pencil lines

Part#1

Ensure flush at
both ends

Glue step#5
Glue part#3 to glue
step#4 assembly

Part#3

Glue step#4
assembly

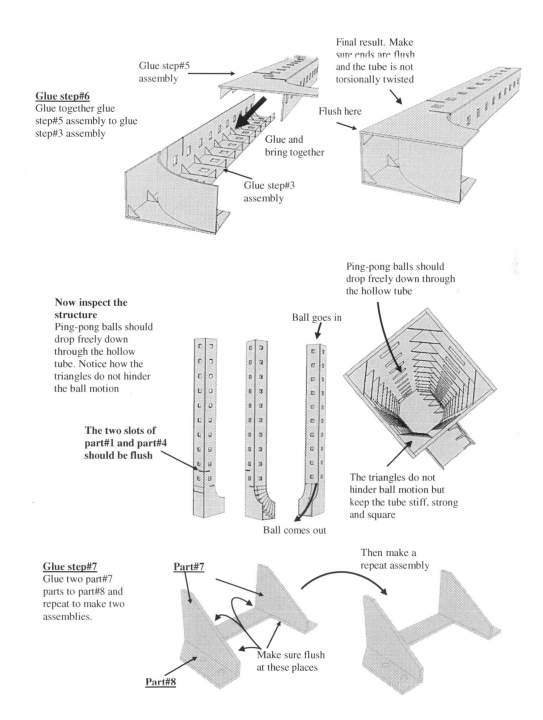

Glue step#5 assembly

Final result. Make sure ends are flush and the tube is not torsionally twisted

Glue step#6
Glue together glue step#5 assembly to glue step#3 assembly

Flush here

Glue and bring together

Glue step#3 assembly

Ping-pong balls should drop freely down through the hollow tube

Now inspect the structure
Ping-pong balls should drop freely down through the hollow tube. Notice how the triangles do not hinder the ball motion

Ball goes in

The two slots of part#1 and part#4 should be flush

The triangles do not hinder ball motion but keep the tube stiff, strong and square

Ball comes out

Then make a repeat assembly

Glue step#7
Glue two part#7 parts to part#8 and repeat to make two assemblies.

Part#7

Make sure flush at these places

Part#8

Now screw these assemblies to S9452 Futaba servos. You need two servos. Use M3 stainless steel screws, 10 long with cross head. Use also stainless steel washers. Self tap the screws into the $\phi 2.5$ mm holes in the cardboard. Make sure the servos have their rubber gaiters fitted. Do not tighten the screws too much because you will strip the self-tap threads. The four screws per servo will hold the servo securely.

Now screw part#9 to the servo horn. You should choose the 21 mm diameter horn as shown. First use four M2 cross head, stainless steel screws 6 mm long to attach part#9 to the horn. Part#9 has a $\phi 7.1$ mm hole which is a registration hole that matches the horn center. This hole also allows you to use a screwdriver to access the screw that secures the horn to the servo shaft. The horn holes

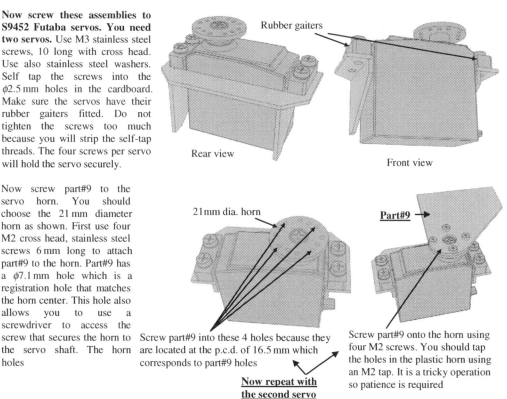

Rubber gaiters

Rear view

Front view

21mm dia. horn

Part#9 →

Screw part#9 into these 4 holes because they are located at the p.c.d. of 16.5 mm which corresponds to part#9 holes

Screw part#9 onto the horn using four M2 screws. You should tap the holes in the plastic horn using an M2 tap. It is a tricky operation so patience is required

Now repeat with the second servo

Now set the horn into its calibrated position. You can refer to Chapter 6 which describes the servo horn calibration

The calibration is to set the horn such that its mid-range position is as shown here. This position will be the "open" position that allows balls to pass through the tube

Gently turn the horn fully CCW to its end stop

Note the orientation of part#8

45°

This angle should be close to 45°. If not remove the horn and replace to give this angle. See Chapter 6 for explanation

Gently turn CW

Gently turn the horn fully CW to its end stop to confirm the orientation is approximately as shown here

Then repeat calibration with second servo. The calibration is the same. There is no "handedness"

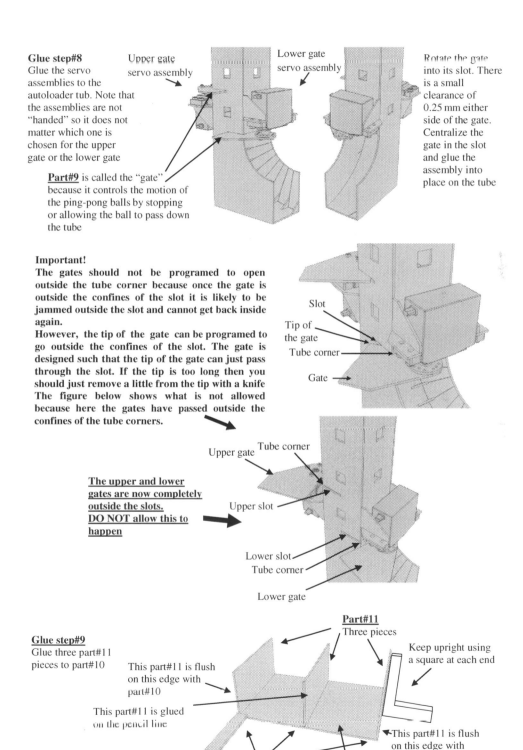

Glue step#8
Glue the servo assemblies to the autoloader tub. Note that the assemblies are not "handed" so it does not matter which one is chosen for the upper gate or the lower gate

Upper gate servo assembly

Lower gate servo assembly

Rotate the gate into its slot. There is a small clearance of 0.25 mm either side of the gate. Centralize the gate in the slot and glue the assembly into place on the tube

Part#9 is called the "gate" because it controls the motion of the ping-pong balls by stopping or allowing the ball to pass down the tube

Important!
The gates should not be programed to open outside the tube corner because once the gate is outside the confines of the slot it is likely to be jammed outside the slot and cannot get back inside again.
However, the tip of the gate can be programed to go outside the confines of the slot. The gate is designed such that the tip of the gate can just pass through the slot. If the tip is too long then you should just remove a little from the tip with a knife The figure below shows what is not allowed because here the gates have passed outside the confines of the tube corners.

Slot
Tip of the gate
Tube corner
Gate

The upper and lower gates are now completely outside the slots. DO NOT allow this to happen

Upper gate   Tube corner
Upper slot
Lower slot
Tube corner
Lower gate

Glue step#9
Glue three part#11 pieces to part#10

Part#11
Three pieces

Keep upright using a square at each end

This part#11 is flush on this edge with part#10

This part#11 is glued on the pencil line

Use the instep jig on all three part#11 pieces

This part#11 is flush on this edge with part#10

Part#10

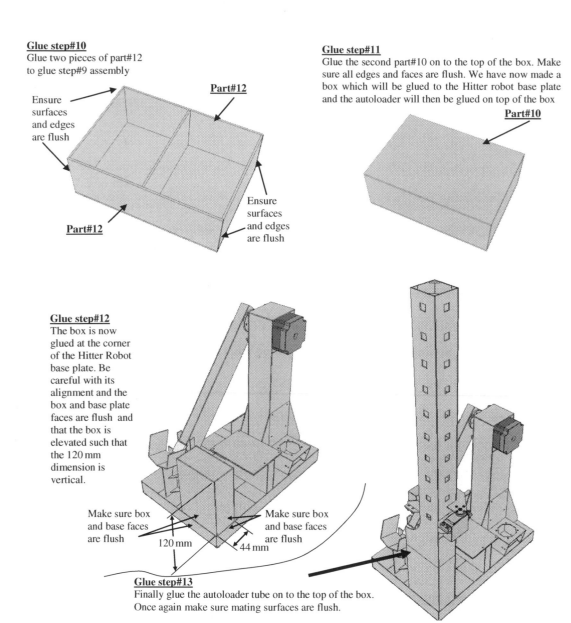

Glue step#10
Glue two pieces of part#12
to glue step#9 assembly

Part#12

Ensure
surfaces
and edges
are flush

Ensure
surfaces
and edges
are flush

Part#12

Glue step#11
Glue the second part#10 on to the top of the box. Make
sure all edges and faces are flush. We have now made a
box which will be glued to the Hitter robot base plate
and the autoloader will then be glued on top of the box

Part#10

Glue step#12
The box is now
glued at the corner
of the Hitter Robot
base plate. Be
careful with its
alignment and the
box and base plate
faces are flush and
that the box is
elevated such that
the 120 mm
dimension is
vertical.

Make sure box
and base faces
are flush

Make sure box
and base faces
are flush

120 mm

44 mm

Glue step#13
Finally glue the autoloader tube on to the top of the box.
Once again make sure mating surfaces are flush.

8.3 **WIRING AND PROGRAMMING THE AUTOLOADER**

The next step is to connect the servos to the Basic Stamp Board of Education and to do testing of code. First of all connect the servos to the Basic Stamp Board using the black servo sockets which are pins 12, 13, 14, and 15. Use pin 12 for the lower servo and pin 13 for the upper servo. See photographs below. Note that the 9 V rechargeable battery shown in the photo is used to drive the servos with the jumper pin set to V_{in}. This means that 9 V is directly connected to the servo. Students should first try to see if the servos work satisfactorily with the jumper set to V_{dd} which means +5 V regulated. Make sure you use rechargeable batteries because (i) it is expensive to keep replacing nonrechargeable batteries and (ii) the internal resistance of rechargeable batteries is lower than nonrechargeable that means the servo current surge when opening and closing gates will not cause brownouts meaning that the battery voltage will stay above 5.5 V which is the minimum for the Basic Stamp to maintain operation. You have to control and calibrate the servos with the PULSOUT instruction such that the pulsout values of each servo in the open and close positions are known. The authors found that the values were as indicated in the PBasic Stamp code that follows.

Upper gate servo
Futaba S9452

Lower gate servo
Futaba S9452

Servo sockets, pin 12 and pin 13. You may have to trim off with a knife the side ridge from the servo plastic plugs to permit entry into the servo sockets

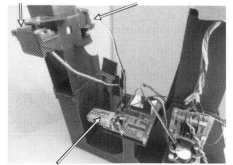

Rechargeable
9 V battery

Basic Stamp Board of
Education fitted with BS2
microcomputer module. The
Board is the USB version

USB
connector

Sample PBasic Stamp Code for Operating the Autoloader That Deposits a Ball Into the Tee

```
'{$STAMP BS2}
'This is the 10-ball magazine autoloader programme
'Remember that when using the BS2 that the clock period is 2µs. If you use the faster BS2px then
'the clock period is 0.81µs. This means that the pulsout instruction will need to be recalibrated.
'More information on servos at chapter 6
lowerpin     CON 12      'control pin number of lower gating servo that gates ball into the tee
upperpin     CON 13      'control pin number of upper gating servo that gates balls before loading,
          'meaning that it holds back the column of balls waiting to be delivered to the tee
upperclose   CON 800   'the pulsout number for the upper gating servo to close gate
upperopen    CON 600   'the pulsout number for the upper gating servo to allow a ball through
lowerclose   CON 790   'pulsout number for closing the lower gating servo
loweropen    CON 1000   'the pulsout number for the lower gating servo to deliver ball to the tee
i            VAR  Word    'variable used for general counting
j            VAR  Word    'variable used for general counting
d            CON  7       'variable that controls ball delivery rate. 5 is fastest

'Programme to deliver 10 balls then stop
          FOR j=1 TO 10   'deliver 10 balls then stop
          GOSUB dropball
          NEXT
again:    GOTO again      'stop the programme

dropball:subroutine for dropping one ball onto the tee
          'drop ball into tee
          FOR i=1 TO d     'variable 'd' represents the time given to change state of the servos
                           'for example, d = 7; time = 7 x (15+(2 x 1.5)approx)=126ms approx
          PULSOUT upperpin,upperclose   'drive upper gate servo to close gate
          PULSOUT lowerpin,loweropen    'drive lower gate servo to drop ball onto the tee
          PAUSE 15                      'update period of the servos, i.e. 15ms
          NEXT

          'close up gates
          FOR i=1 TO d     'same time given to change stet of servos
          PULSOUT upperpin,upperclose   'keep the upper gate closed, (no change from previous state)
          PULSOUT lowerpin,lowerclose   'close the ball delivery lower gate
          PAUSE 15
          NEXT

          'drop next ball into next ball cavity ready for delivery to tee
          FOR i=1 TO d
          PULSOUT upperpin,upperopen   'allow in a new ball into the delivery cavity
          PULSOUT lowerpin,lowerclose   'keep the lower gate closed, (no change from previous state)
          PAUSE 15
          NEXT
```

```
'close up gates
FOR i =1 TO u
PULSOUT upperpin,upperclose   'close up gate to hld column of balls in check
PULSOUT lowerpin,lowerclose   'keep the lower gate closed, (no change from previous state)
PAUSE 15
NEXT
RETURN       'end of delivery of one ball programme, return to main programme
```

8.4 **PROBLEMS**

1. Fit a ball sensor that detects that a ball is sitting in the tee. Sometimes there is a fault and a ball is not delivered because of a jam or because the ball bounces out of the tee.

2. Experiment to find out the maximum delivery rate of balls in balls per second.

3. Write a program for the Hitter robot that maintains a constant speed of the hitting arm for rapid fire. In other words, the hitting arm does not stop; it just keeps going. Hitting balls that drop onto the tee. The maximum delivery rate of the autoloader is probably two balls per second so it will take 5s to empty the magazine. The maximum arm rotational velocity is probably 4 revs/s. This means that the hitting arm can only hit a ball every alternate revolution.

4. Working with a classmate, hit 10 balls and mark on the floor where the balls touchdown. Now analyze the touchdown variation. Try to work out why there is a grouping, in other words what causes these repeatability errors. Try to find ways to reduce the repeatability errors.

5. The stepper motor together with its driver introduces vibration to the tee which causes the ping-pong ball to tremble in its three-point mounting. This may cause hitting distance repeatability errors. Examine possible solutions to this problem.

6. How can the Hitter robot be redesigned to vary the ball launching angle?

7. The hitter arm can be a safety problem. Design a safety guard or a smart proximity sensing system that deactivates the arm if a person is too close to the robot.

Theory VI: Theory and Design Notes Related to the Throwing Robot

LEARNING OUTCOMES

1. Dynamical force analysis.
2. Kinematic analysis of a mechanism.
3. Servo-actuated mechanism design.

9.1 WORKING CONCEPT OF THE THROWING ROBOT

The Throwing robot is designed to throw a ping-pong ball up to 6 m range. The idea behind the throwing mechanism is to hold the ping-pong ball in a whirling end effector, like the arm and hand of a cricket bowler, and let go the ball at a precise angle and speed to throw accurately the ball (Fig. 9.1). An end effector is an elementary robot hand with a finger or claw.

This end effector is mounted at the extremity of the whirling arm which is rotated by a stepper motor that is mounted on top of a tower structure. The arm rotates in a vertical plane and releases the claw via a servo-actuated link that opens the claw to release the ball when it achieves a precise launch velocity, **v**. The whirling arm is known as a "smart arm" because of its computationally intelligent onboard microcomputer that computes the precise angle to release the ball. The stepper motor, driven by a second separate microcomputer, is responsible for producing the precise launch speed. Thus, working synergistically together, the stepper motor system and the smart arm system produce a precise ball launch velocity.

9.2 BASIC DESIGN REQUIREMENTS OF THE THROWING ROBOT

The Excel spreadsheet analysis in Chapter 3 showed that an approximate ball launch speed of 12 m/s is required to achieve a throwing distance of 6 m. The maximum achievable angular velocity of the stepper motor using a 24 V power supply with the Longshine DM542 stepper motor driver,

Creating Precision Robots. https://doi.org/10.1016/B978-0-12-815758-9.00009-2

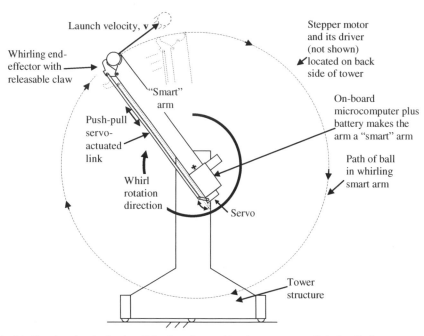

■ FIG. 9.1 Principle of the Throwing robot showing the whirling "smart" arm and its microcomputer controlled releasable claw.

Chapter 7, was found to be 5.5 rev/s (=34.6 rad/s), thus, to achieve, 12 m/s launch speed, *the ball must be held in the releasable claw at a radius,* $r = v/\omega = 12/34.6 = 0.35 m$.

9.3 DESIGN OF THE CLAW AND ITS RELEASABLE MECHANISM

The claw and its releasable mechanism have been carefully designed. First of all, similar to the Hitter robot tee, the ball is mounted on three points, known as a nonredundant kinematic mounting system that restricts displacement of the ball in four directions, which are, $\pm x$, $-y$, and $-z$ directions according to Fig. 9.2. The captive force presses the ball gently against the three points and restricts ball motion in the remaining two directions, that is, $+y$ and $+z$, hence the ball is constrained fully in x, y, and z directions. Fig. 9.3 is a view in the x-z plane where it can be seen that points a and b are separated by 30 degrees either side of the z-axis. The angles could be more and they could be less but ±30 degrees was decided upon as suitable values. Fig. 9.4 shows a side view of the arm in the y-z plane. Here, it is important to note that the claw relies on an elastic band to keep it latched closed without servo actuation, providing the ball captive force and to keep it open also without servo actuation. Servo actuation is only used for a short impulsive

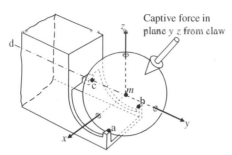

■ **FIG. 9.2** Three-point nonredundant kinematic ball mounting system at arm extremity. Ball is located at points a, b, and c. Ball mass center (coincident with volume center) is at point *m*. The *x*, *y*, and *z* axes originate at point *m*.

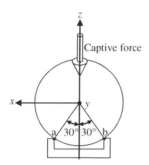

■ **FIG. 9.3** Principle of ball mounting in the *x-z* plane.

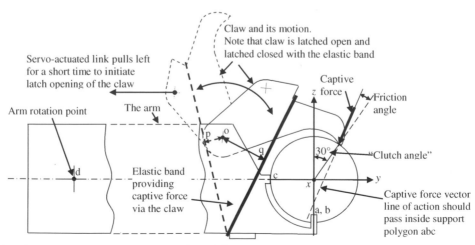

■ **FIG. 9.4** Principle of ball mounting in the *y-z* plane. Elastic bands apply ball captive force via the claw. Note that the *y*-axis, projected to the left, passes through point d, the arm rotation point.

force either to latch the claw closed or to latch it open. Once the claw is either open or closed then the servo is switched off.

The latch works like a household electric light switch which is a mechanical spring device. You press the switch in one direction to switch the light on and the switch "latches," that is, stays, in that position, then you switch it in the other direction and it latches in that new position and the light stays off.

The latch works due to the elastic band moment arms, op and oq in Fig. 9.4 creating a CCW and CW moment on the claw causing the claw to latch open and to latch closed, respectively. In other words, the claw uses an elastic band to apply a captive force on the ball in the closed position and, the same elastic band keeps the claw open. So, a servo and a link, together with the elastic band, create a two-state switchable mechanical system that is analogous to a bistable flip-flop in digital electronics.

Note that the direction of the captive force in Fig. 9.4 is not entirely known due to the friction angle between the claw and the ball. The figure shows that the claw tip presses on a point on the ball that is in the plane y-z and subtends an angle of 30 degrees from the z-axis. This angle is called the "clutch angle." The angle could be 0 degree but the ball may slip out and the ball may not be pressed against point c. If the angle is 90 degrees, then the ball may not be pressed against points a and b and furthermore the claw may not open fast enough to release the ball unimpeded. So, a suitable compromise was to set the angle to 30 degrees but even this angle may mean that the captive force vector line of action may fall outside the ball support polygon, a, b, c. Anyway, this is uncertainty represents sometimes the difficulties that designers have to face. An good test to check the effectiveness of the end effector is to gently pull out the ball a small distance, say 5 mm, in the y-direction and then to let go. If the captive force vector line of action passes inside the support polygon, then the ball will be "snapped in," that is, pulled in to the clutch of the end effector. In other words, you would not have to push it all the way into the three-point mounting. If this test does not work, then you will have to remake the claw such that clutch angle is greater than 30 degrees, maybe 40 degrees.

9.4 **BALL RELEASE KINEMATIC ANALYSIS**

Fig. 9.5 shows a proposed equal spaced time-lapse diagram of the ball leaving the end effector in order to examine the ball trajectory immediately on leaving the end effector after being released from its three-point mounting. It is important to examine the immediate leaving trajectory because the claw and three-point mounting can corrupt the precision of the launch velocity.

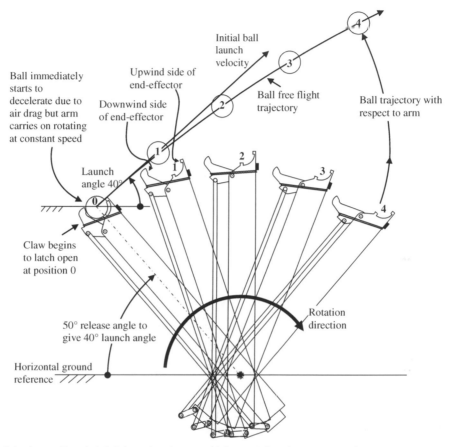

Initial ball
launch
velocity

Upwind side of
end-effector

Ball immediately
starts to
decelerate due to
air drag but arm
carries on rotating
at constant speed

Downwind side
of end-effector

Ball free flight
trajectory

Ball trajectory with
respect to arm

Launch
angle 40°

Claw begins
to latch open
at position 0

Rotation
direction

50° release angle to
give 40° launch angle

Horizontal ground
reference

■ **FIG. 9.5** Equal time-lapse positions, 0, 1, 2, 3 4,... of rotating arm and ball, respectively, *with respect to ground.*

For a high-precision launch velocity, the claw tip should move away from the ball sufficiently fast so as not to stay in lingering contact with the ball nor should the tip hit the ball as the ball exits the end effector. The same goes for the three-point mounting; there should be a clean lift off of the ball from the three-point mounting with no further contact with the ball. To enable these actions, it is important to note that the claw is on the downwind side of the ball throwing direction and the three-point mounting is on the upwind side. This was done deliberately because as soon as the ball is released from the three-point mounting, the air drag will start to decelerate the ball. Thus, the ball will directly and cleanly lift off from the three-point mounting rather than be pressed into the mounting. This would happen if the claw were to be on the upwind side of the ball throwing direction. The claw may now be a problem so in order for that not to be so, the claw must move away fast enough from the ball so as not to interfere with the ball flight.

For example, Fig. 9.5 shows the ball trajectory relative to the arm, not relative to the ground as in Fig. 9.4, on being released from the end effector. The angle with respect to the arm x-axis of the ball trajectory at the point of ball release is θ_{exit} and the initial speed with respect to the arm would be expected to be zero because the ball has been traveling captively with the end effector. However, we are not in a position to judge the exit angle, θ_{exit}, nor whether the claw release velocity fast is enough so as not to interfere with the ball trajectory. A high-speed camera would be helpful but otherwise it's a good area for students to do some research to investigate what really happens at ball release (Fig. 9.6).

9.5 FORCES ACTING ON PING-PONG BALL DURING ROTATION

Fig. 9.7 shows the Throwing robot smart arm at two positions, which are, generalized angle, θ and specific angle, $\theta = 270$ degrees during its rotation. Four accelerations are acting on the ball, namely (i) centripetal, (ii) Coriolis, (iii) tangential, and (iv) radial. The Coriolis and radial accelerations are zero because the ball is restricted from radial motion. The ball experiences the remaining two accelerations which are, centripetal and tangential but of these two, only the centripetal acceleration will try to extract the ball from the clutch of the end effector. Thus, we only need to consider the centripetal acceleration in our calculations together with the ball weight when the arm is

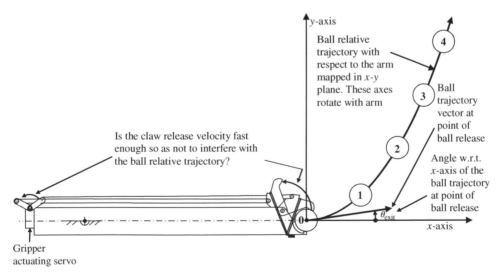

■ **FIG. 9.6** The same time-lapse positions of the ball, 0, 1, 2, 3, 4,... but this time with respect to the arm. The ball release mechanism must be fast enough so as not to interfere with the trajectory of the ball immediately after its release.

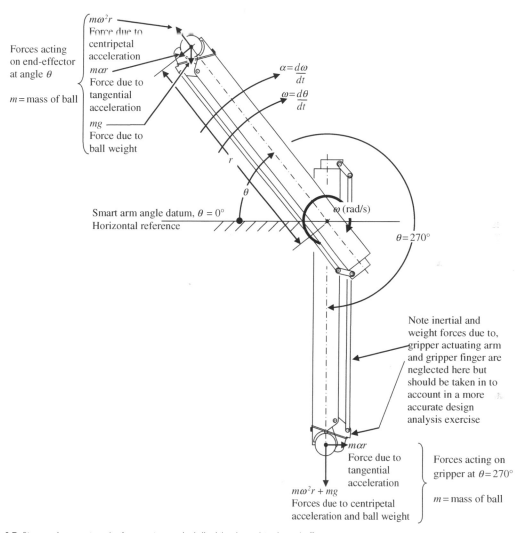

Forces acting
on end-effector
at angle θ

m = mass of ball

$\begin{cases} m\omega^2 r \\ \text{Force due to} \\ \text{centripetal} \\ \text{acceleration} \\ m\alpha r \\ \text{Force due to} \\ \text{tangential} \\ \text{acceleration} \\ mg \\ \text{Force due to} \\ \text{ball weight} \end{cases}$

$\alpha = \dfrac{d\omega}{dt}$

$\omega = \dfrac{d\theta}{dt}$

r

θ

Smart arm angle datum, $\theta = 0°$
Horizontal reference

ω (rad/s)

$\theta = 270°$

Note inertial and
weight forces due to,
gripper actuating arm
and gripper finger are
neglected here but
should be taken in to
account in a more
accurate design
analysis exercise

$m\alpha r$
Force due to
tangential
acceleration

$m\omega^2 r + mg$
Forces due to centripetal
acceleration and ball weight

$\begin{cases} \text{Forces acting on} \\ \text{gripper at } \theta = 270° \\ \\ m = \text{mass of ball} \end{cases}$

■ **FIG. 9.7** Diagram that examines the forces acting on the ball whilst clamped in the end effector.

vertically downwards. The arm in this position is when ball weight tries to extract the ball from the end effector.

The positive direction of angle, θ, angular velocity, $\omega = d\theta/dt$, and angular acceleration, $\alpha = d\omega/dt$ are all shown in the CW direction.

The two forces acting on the three-point mounting and claw are (i) the weight of the ball, mg and (ii) the centrifugal force of the ball, $m\omega^2 r$. The directions of each of these forces combine to form a maximum at the vertical downwards direction, that is, $\theta = 270$ degrees.

The forces are calculated from the following data:

Ball mass, $m = 2.7\,$g measured and in agreement with internet specification

Maximum angular velocity, $\omega = 34.6\,$rad/s $\approx 35\,$rad/s

Approximate gravitational acceleration, $g = 10\,$m/s^2

Ball center of gravity radius from center of rotation, $r = 0.35\,$m

Thus, approximate magnitude of ball forces acting on end effector are:

Ball centripetal acceleration $= \omega^2 r = 35^2 \times 0.35 = 430\,$m/s^2
(Wow! ball will experience 43 times its static weight)

Thus, centrifugal force, $F_c = m\omega^2 r = 0.0027\,$kg $\times 430\,$m/s$^2 = 1.16\,$N (three sig. figs. and thus two dec.pl.)

Ball weight, $w_{ball} = mg = 0.0027 \times 10 = 0.027\,$N $= 0.03\,$N (two dec. pl. to match the centrifugal force precision)

It can be seen that the centrifugal force dominates the ball weight. The worst case arm position for the likelihood of the ball being wrenched out of the end effector is when $\theta = 270$ degrees, that is, the arm is vertically downwards as in Fig. 9.7. Here, the gravitational and centrifugal forces align and sum together, that is, $0.03\,$N $+ 1.16\,$N $= 1.2\,$N (two sig. figs.).

In Conclusion

There will be a maximum radial force of 1.2 N trying to pull the ball out from the clutch of the end effector in the radial direction.

9.6 METHOD FOR CHECKING ADEQUACY OF END-EFFECTOR CLUTCHING FORCE

The estimated maximum force that is pulling the ball out of the clutch of the end effector has just been calculated as 1.2 N and this force is radially outwards from the center of rotation of the arm. If the end-effector ball clutching force is inadequate, then during the arm whirling, the ball will fly out from the end effector prematurely before the end-effector release mechanism is activated. We can check adequacy of end-effector ball clutching force, simply, by using a spring weigh scale as in Fig. 9.8. As a factor of safety, elastic bands should be added until the end-effector ball clutching force exceeds 1.5 N or maybe 2.0 N. Bear in mind that if you write software to increase the arm angular velocity then the centrifugal force increases proportional to the square of angular velocity. Note that structural vibrations and resonances induced by the stepper motor may cause premature release of the ball even though the clutching force has been set to 1.5 N. In this case, the clutching force should be increased. It should be noted that the link will add or subtract an additional gripping force due to its dynamical forces

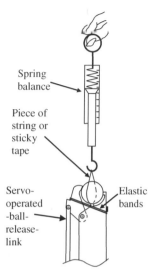

Spring balance

Piece of string or sticky tape

Servo-operated-ball-release-link

Elastic bands

■ **FIG. 9.8** Method of checking adequacy of the ball pull out force using a spring balance and piece of string tied around the ball. Note that the servo-operated-ball-release link is deactivated. Thus, the ball is solely restrained in its location by the force of the elastic bands.

acting on the claw. This is an interesting exercise. Students can calculate this force by measuring its mass and position of center of gravity together with its location in the mechanical system.

9.7 **REPEATABILITY ERRORS**

The concept of repeatability errors is one of the most intriguing and perplexing difficulties in engineering product design. It is also, arguably, one of the most important concepts since it is related to product performance and cost and hence whether the product is commercial or not.

By now, students will have been throwing or hitting ping-pong balls many times and it will be very apparent that the balls land at different points every time. Why is this? The computer program was the same, different balls may have been used but they are similarly manufactured balls aren't they? Even if the same ball is used it is still found that the ball lands in a different place every time. What students are seeing is a firsthand experience of engineering repeatability errors. It exists whenever products are manufactured and the causes of repeatability errors are usually far from obvious. Human beings are very used to seeing repeatability errors that we do not ask why they exist. For example, in sport, why cannot we pocket the basketball in the basket every time, or why in tenpin bowling cannot we hit a strike-10 every time. Machines that we manufacture also give different results every time. The decreasing of repeatability errors is allied with an increasing quality and sometimes, but not always, increasing cost. Engineering is all about ingenuity and many times if you are ingenious then you can increase quality and decrease repeatability errors at less cost, not more cost. To do this requires careful thought and deeper understanding of the physics of the engineered product. One job of research, development and design, (RD&D) engineers is:

(i)   to identify the source of repeatability errors,
(ii) to rank them in order of greatest effect on errors, and then
(iii) find ways of reducing these errors

Typical methods of reducing repeatability errors are:

(i)   Tolerancing components.
(ii) Changing the design of the product.
(iii) Using active compensation.
(iv) Improving the manufacturing procedures.
(v)   Using alternative materials or components.
(vi) Improving the computer control algorithm.

Active compensation means that, for example, if the temperature of the environment causes an increase of ball throwing distance of 10 cm/°C, then the repeatability error can be reduced by measuring temperature and modifying the ball launch speed accordingly.

After doing numerous hitting or throwing tests, students should write down what they think causes repeatability errors in terms of the ball landing point. This means not just the distance along the trajectory path but also the distance orthogonal to the trajectory path, that is, to the left and right of the point of landing; in other words, what causes the ball to swerve left or right along its trajectory? Do not worry if any causes of repeatability error seem ridiculous; just write down anything and everything that comes to mind. Once you have a repeatability error source list then rank these in descending order from your feeling of greatest effect on error to minimum effect on error. Of course, you do not know if the order is correct but its a start. Now, set about canceling the effect of each of these causes. For example, if you think that the lack of balance of the smart arm is a culprit, then set about balancing the arm. An out-of-balance rotating arm will vibrate the tower structure and may lead to varying launch angles. Or, maybe the tower structure is not stiff enough and its vibrations will cause errors. You can stiffen the structure with "wing" stiffeners (Fig. 9.9).

9.8 **PROBLEMS**

1. What is the effect of the ball component of rotation, equal to the arm angular velocity, as it leaves the end effector. Does the Magnus effect change its trajectory?

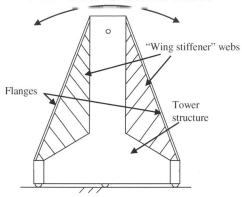

■ **FIG. 9.9** Decreasing repeatability errors via wing stiffeners.

2. Work out the *x*-axis and *y*-axis component of position with respect to time for a ball with zero air drag. Then, add the air drag component based on data in Chapter 3 to give the deceleration of the ball and then integrate twice with respect to time to get an additional air drag displacement. You should be in a position now to work out what the claw opening with

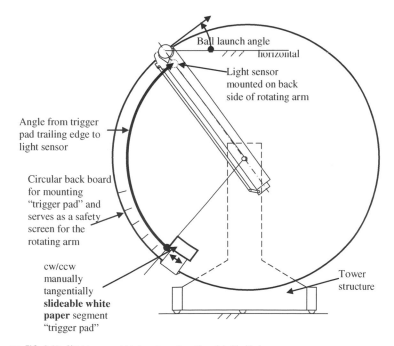

■ **FIG. 9.10** Obtaining a variable launch angle with a slideable block.

respect to time should be to clear itself from the ball. Problem is now how to measure the claw opening with respect to time.

3. Calculate dynamical forces on the claw due to the servo-actuated link system.

4. Obtain a variable angle of launch with a slideable white paper block on an arc (Fig. 9.10).

The Ball Throwing Robot: Design and Construction

LEARNING OUTCOMES

1. Experience in building lightweight, stiff, and strong mechanical structures and mechanisms.
2. Knowledge of a more advanced mechatronic system.
3. Parallel real-time programming experience of a dual core microcomputer system.
4. Method of impulse programming of an integrated servomechanism as a bidirectional mechanical latch.
5. Using a sensor to give accurate datuming.

The Throwing robot is split into two separate building entities which are (i) the tower structure and (ii) the smart arm. We start with the tower structure which houses the stepper motor and that is similar to the tower structure of the Hitter robot but taller. Once again all parts are made from 1.5 mm thick cardboard.

10.1 THE TOWER STRUCTURE (ALL DIMENSIONS IN MM)

That's the end of the tower construction. Now we move on to the smart arm. Here are its drawings. Part numbers continue from the tower part numbers.

Creating Precision Robots. https://doi.org/10.1016/B978-0-12-815758-9.00010-9

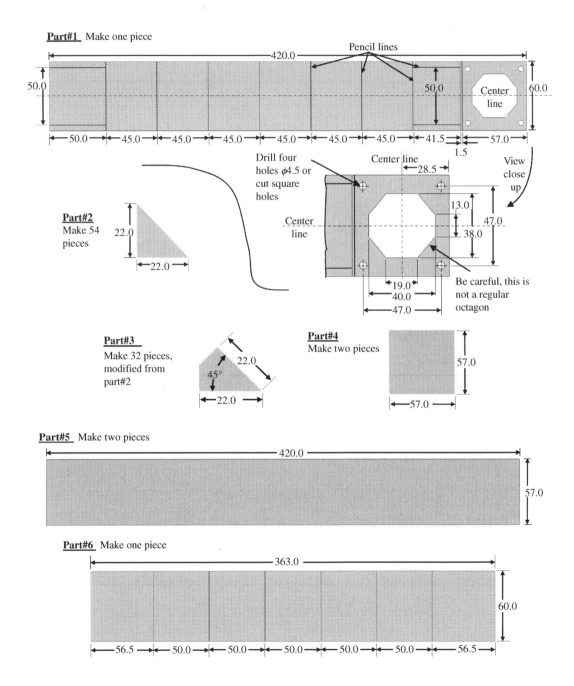

Part#1 Make one piece

Pencil lines

420.0

50.0

50.0

60.0

Center line

50.0 — 45.0 — 45.0 — 45.0 — 45.0 — 45.0 — 45.0 — 41.5 — 57.0

1.5

Drill four holes φ4.5 or cut square holes

Center line

28.5

View close up

Part#2
Make 54 pieces

22.0

22.0

Center line

13.0

47.0

38.0

19.0

40.0

47.0

Be careful, this is not a regular octagon

Part#3
Make 32 pieces, modified from part#2

22.0

45°

22.0

Part#4
Make two pieces

57.0

57.0

Part#5 Make two pieces

420.0

57.0

Part#6 Make one piece

363.0

60.0

56.5 — 50.0 — 50.0 — 50.0 — 50.0 — 50.0 — 56.5

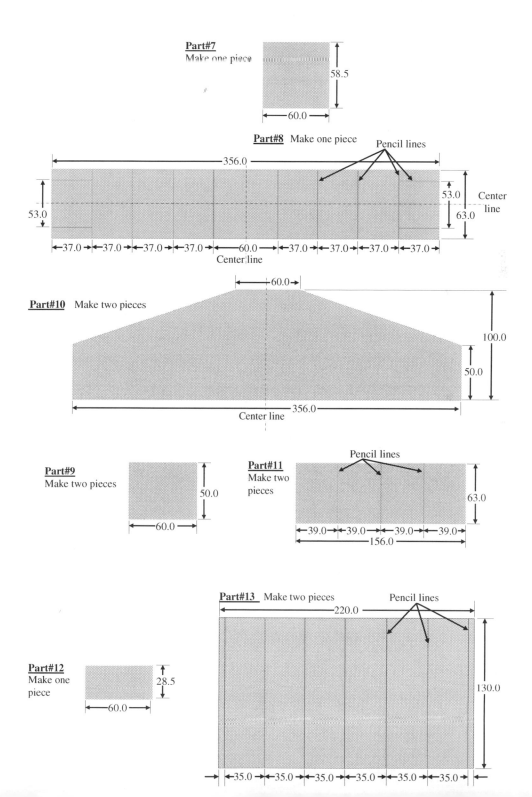

Part#7
Make one piece

58.5

60.0

Part#8 Make one piece   Pencil lines

356.0

53.0
Center line

53.0

63.0

37.0 37.0 37.0 37.0 60.0 37.0 37.0 37.0 37.0
Center line

60.0

Part#10 Make two pieces

100.0

50.0

356.0
Center line

Part#9
Make two pieces

50.0

60.0

Part#11
Make two
pieces

Pencil lines

63.0

39.0 39.0 39.0 39.0
156.0

Part#13 Make two pieces   Pencil lines

220.0

130.0

Part#12
Make one
piece

28.5

60.0

35.0 35.0 35.0 35.0 35.0 35.0

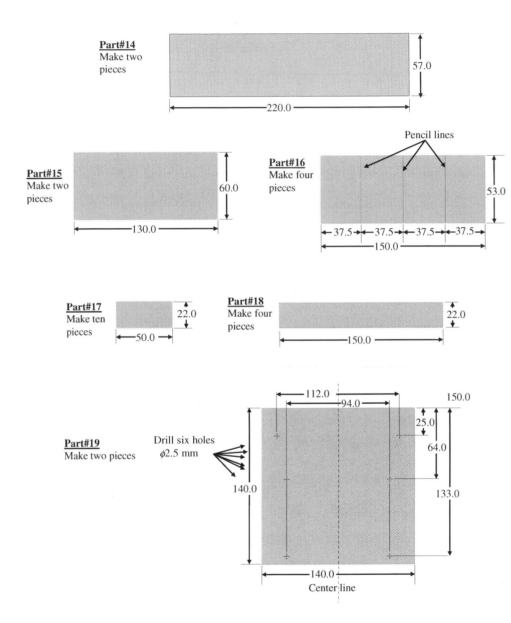

Now glue up the parts for the Tower

Glue step#1

Glue 18 pieces of **part#2** to **part#1** using the instep jig. All triangles to be glued on the pencil lines

Instep jig to position parts 1.5 mm inboard of edge of part#1

These two triangles to be glued 1.5 mm from edge

Part#2 18 pieces

Part#1

These two triangles to be placed up to pencil line (first line not second line)

Glue step#2

Glue two pieces of **part#4** to previous glue step#1 assembly using instep jig

Part#4

Make sure these parts are abutted up against the triangles and 1.5 mm inboard from edge of part#1

Part#1

Glue step#3

Glue two pieces of **part#5** to previous glue step#2. Temporarily put this assembly aside whilst doing the next glue step

Part#5

Make sure part#5 is glued abutted against the triangles

Make sure edges and faces are flush

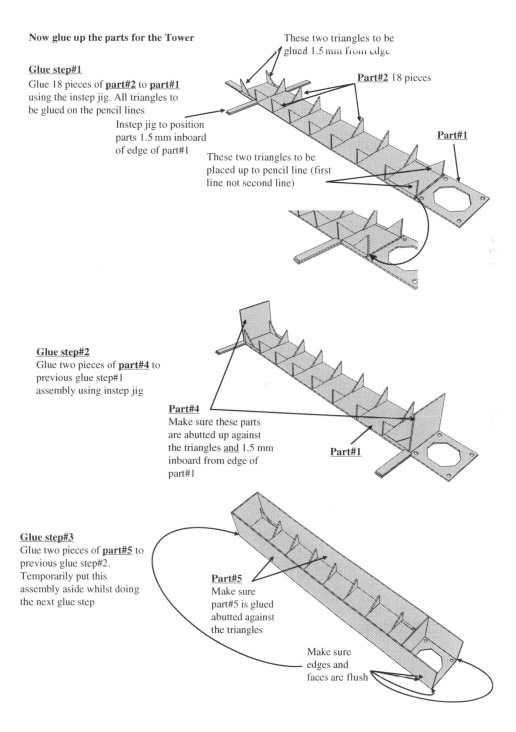

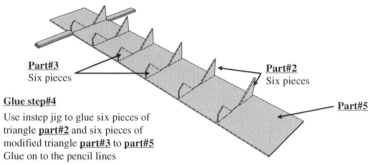

Part#3
Six pieces

Part#2
Six pieces

Part#5

Glue step#4

Use instep jig to glue six pieces of
triangle **part#2** and six pieces of
modified triangle **part#3** to **part#5**
Glue on to the pencil lines

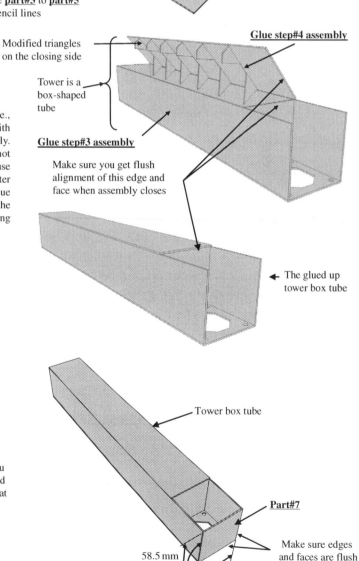

Modified triangles
on the closing side

Glue step#4 assembly

Tower is a
box-shaped
tube

Glue step#5

Close the tower box tube, i.e.,
glue step#3 assembly, with
the glue step#4 assembly.
Make sure that the tube is not
twisted before gluing because
it cannot be untwisted after
gluing. Snap shut the glue
step#4 assembly with the
modified triangles on closing
side as shown

Glue step#3 assembly

Make sure you get flush
alignment of this edge and
face when assembly closes

← The glued up
tower box tube

Tower box tube

Glue step#6

Finish the tower box tube by
gluing on part#7. Make sure you
correctly match the 58.5 mm and
the 60.0 mm dimensions and that
the edges and faces are flush

Part#7

Make sure edges
and faces are flush

58.5 mm

60.0 mm

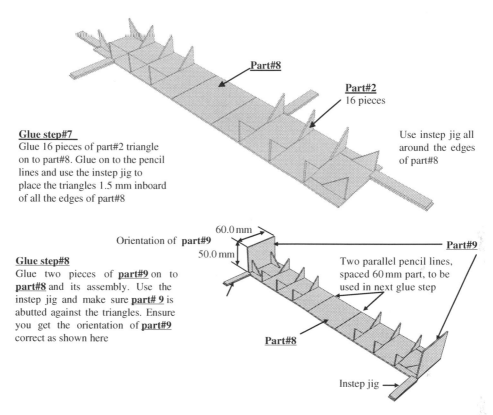

Part#8

Part#2
16 pieces

Use instep jig all
around the edges
of part#8

Glue step#7
Glue 16 pieces of part#2 triangle
on to part#8. Glue on to the pencil
lines and use the instep jig to
place the triangles 1.5 mm inboard
of all the edges of part#8

Orientation of **part#9**

60.0 mm

50.0 mm

Part#9

Two parallel pencil lines,
spaced 60 mm part, to be
used in next glue step

Glue step#8
Glue two pieces of **part#9** on to
part#8 and its assembly. Use the
instep jig and make sure **part# 9** is
abutted against the triangles. Ensure
you get the orientation of **part#9**
correct as shown here

Part#8

Instep jig

Glue step#9
Glue the Tower
box tube between
the two parallel
pencil lines on
part#8 that are
spaced 60 mm apart,
(refer to previous
glue step#8). Use
two instep jigs as
shown. Ensure that
the tower box tube
is at 90° to part#8.
Use an accurately
cut piece of
cardboard as a jig
to do this

Tower
box tube

Tower
box tube

Part#8

Part#10

Instep jigs

Part#10

Glue step#10
Glue two pieces of part#10
on to the sides of the tower
box and on to the previous
glue assembly

Ensure edges and faces are
flush and part#10 is abutted
against the triangles

Glue step#11
Glue six pieces of part#3 on to the pencil lines of **part#11** using the instep jig. You need to make two assemblies

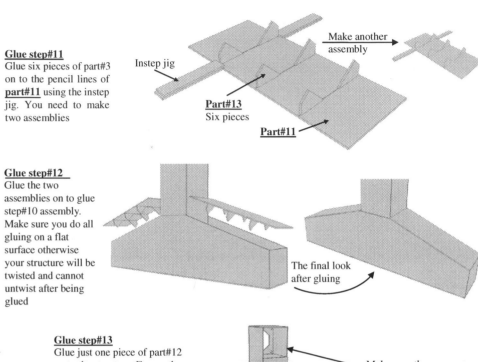

Instep jig

Make another assembly

Part#13
Six pieces

Part#11

Glue step#12
Glue the two assemblies on to glue step#10 assembly. Make sure you do all gluing on a flat surface otherwise your structure will be twisted and cannot untwist after being glued

The final look after gluing

Glue step#13
Glue just one piece of part#12 on to the structure. Ensure the correct orientation as shown and that edges are and faces are flush

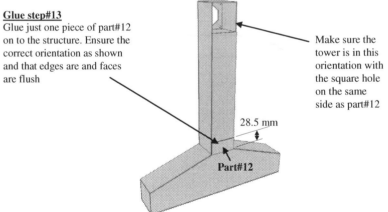

Make sure the tower is in this orientation with the square hole on the same side as part#12

28.5 mm

Part#12

Glue step#15
Glue two pieces of **part#14** on to the previous glue
assembly#14. Ensure faces and edges are flush.
When done, put aside until completion of the next
glue step#16

Glue step#14
Glue 14 pieces of
part#2 triangle on to
part#13 pencil lines
and set 1.5 mm
inboard using the
instep jig

Part#2
14 pieces

Part#13

Instep
jig

Instep jig

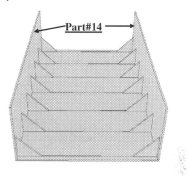

Part#14

Glue step#16
Glue 14 pieces of **part# 3** on to another **part#13**. Once
again glue on to the pencil lines and use the instep jig

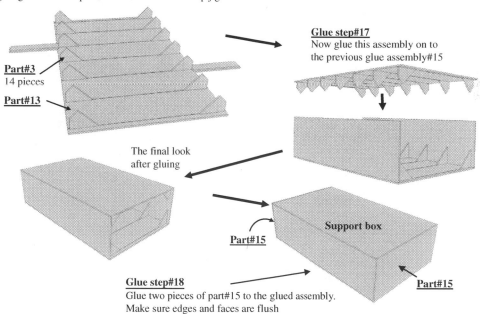

Part#3
14 pieces

Part#13

Glue step#17
Now glue this assembly on to
the previous glue assembly#15

The final look
after gluing

Support box

Part#15

Part#15

Glue step#18
Glue two pieces of part#15 to the glued assembly.
Make sure edges and faces are flush

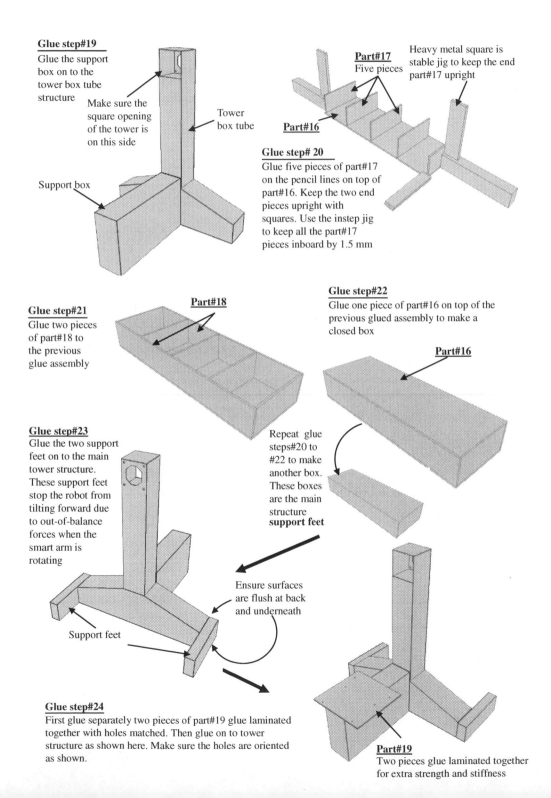

Glue step#19
Glue the support box on to the tower box tube structure

Make sure the square opening of the tower is on this side

Tower box tube

Support box

Part#17
Five pieces

Heavy metal square is stable jig to keep the end part#17 upright

Part#16

Glue step# 20
Glue five pieces of part#17 on the pencil lines on top of part#16. Keep the two end pieces upright with squares. Use the instep jig to keep all the part#17 pieces inboard by 1.5 mm

Glue step#21
Glue two pieces of part#18 to the previous glue assembly

Part#18

Glue step#22
Glue one piece of part#16 on top of the previous glued assembly to make a closed box

Part#16

Glue step#23
Glue the two support feet on to the main tower structure. These support feet stop the robot from tilting forward due to out-of-balance forces when the smart arm is rotating

Repeat glue steps#20 to #22 to make another box. These boxes are the main structure **support feet**

Support feet

Ensure surfaces are flush at back and underneath

Glue step#24
First glue separately two pieces of part#19 glue laminated together with holes matched. Then glue on to tower structure as shown here. Make sure the holes are oriented as shown.

Part#19
Two pieces glue laminated together for extra strength and stiffness

The laminated parts#19 should look like this when the tower structure is viewed from above

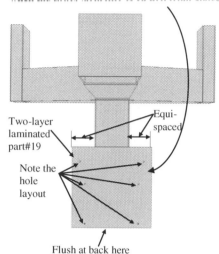

Two-layer laminated part#19

Equi-spaced

Note the hole layout

Flush at back here

Now, add four pieces of stick-on rubber feet. Locate them at the corners of the bottom surface of the structure. You can buy these feet from various sources, e.g., Parallax.com

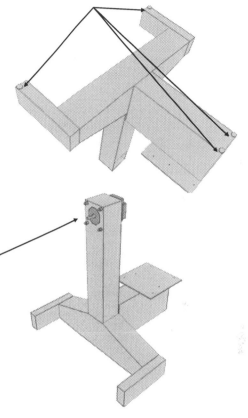

...and finally, complete the Tower structure by adding the stepper motor. Follow the instructions in Chapter 6 which describes the attachment of the stepper motor in the Hitter robot. From now on, when you pick up the robot, you should grip it by the stepper motor because that is, by far, the heaviest component of the robot. If you hold the robot at any other location it will be (i) awkward and unwieldy to do so and (ii) may overstress the structure and cause damage.

10.2 **THE SMART ARM**

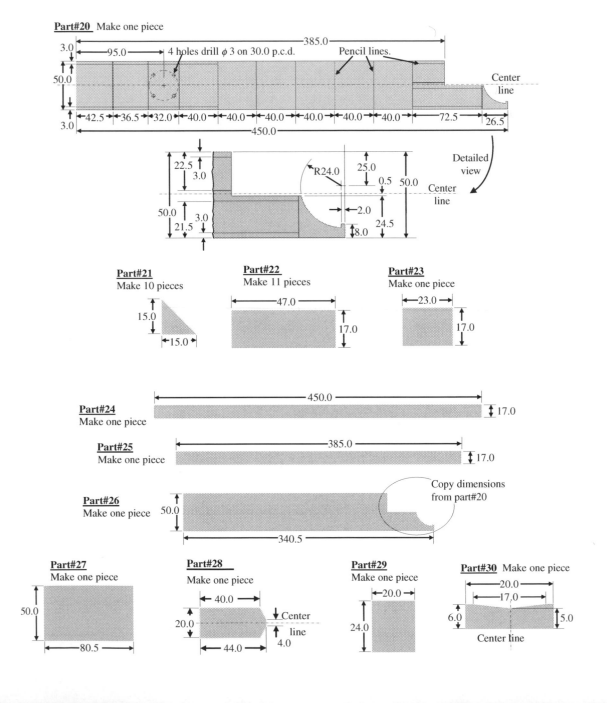

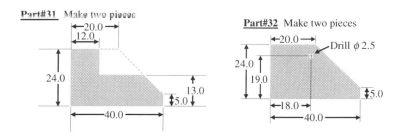

Part#31 Make two pieces

Part#32 Make two pieces

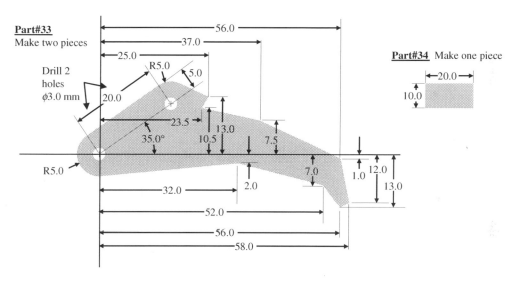

Part#33
Make two pieces

Part#34 Make one piece

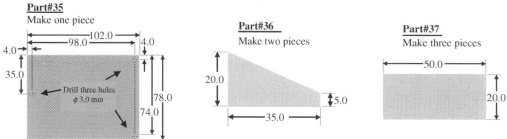

Part#35
Make one piece

Part#36
Make two pieces

Part#37
Make three pieces

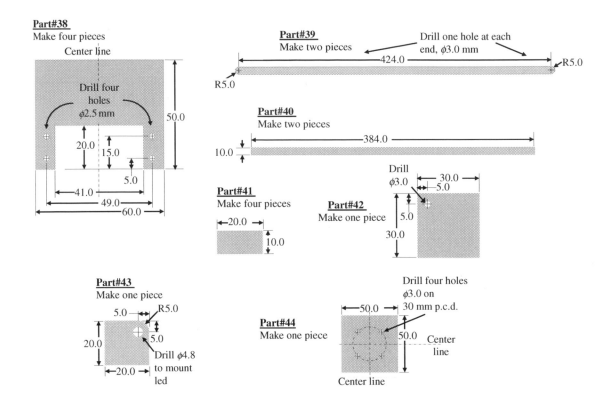

Part#38
Make four pieces
Center line
Drill four holes
ϕ2.5 mm
50.0
20.0
15.0
5.0
41.0
49.0
60.0

Part#39
Make two pieces
Drill one hole at each end, ϕ3.0 mm
424.0
R5.0
R5.0

Part#40
Make two pieces
384.0
10.0

Part#41
Make four pieces
20.0
10.0

Part#42
Make one piece
Drill ϕ3.0
30.0
5.0
5.0
30.0

Part#43
Make one piece
5.0
R5.0
20.0
5.0
Drill ϕ4.8 to mount led
20.0

Part#44
Make one piece
Drill four holes ϕ3.0 on 30 mm p.c.d.
50.0
50.0
Center line
Center line

That's the end of the smart arm drawings so let us get on with building it.

Smart Arm Construction

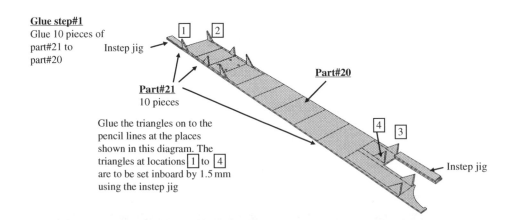

Glue step#1
Glue 10 pieces of part#21 to part#20

Instep jig

Part#21
10 pieces

Part#20

Glue the triangles on to the pencil lines at the places shown in this diagram. The triangles at locations ①to ④ are to be set inboard by 1.5 mm using the instep jig

Instep jig

Glue step#2

Glue 11 pieces of part#22 and one piece of part#23 to part#20 of the glued assembly

Instep jig
Use for all 12 pieces along this edge

Five of the part#22 pieces are abutted against the part#21 triangles. The remaining six pieces of part#22 should be glued upright by eye on to the pencil lines

Part#22
11 pieces

Part#23

Part#20

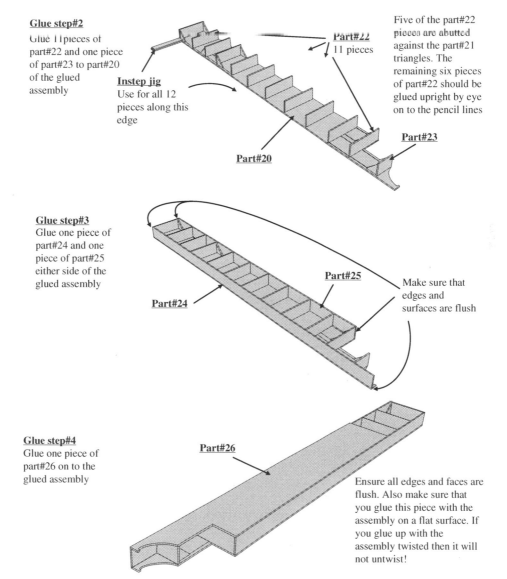

Glue step#3
Glue one piece of part#24 and one piece of part#25 either side of the glued assembly

Part#24

Part#25

Make sure that edges and surfaces are flush

Glue step#4
Glue one piece of part#26 on to the glued assembly

Part#26

Ensure all edges and faces are flush. Also make sure that you glue this piece with the assembly on a flat surface. If you glue up with the assembly twisted then it will not untwist!

The smart arm is now largely a "closed box" except for the open section that is used to access the screws to attach to the stepper motor flange. The closed box means that the arm is lightweight, and stiff and strong in bending and torsion. The open section is kept stiff and strong by virtue of being attached to the flange

Open section for screwing arm to the stepper motor flange

Part#27

Once again make sure all edges and faces are flush

Glue step#5
Glue one piece of part#27 to the assembly

Glue step#6
Glue one piece of part#28 to the assembly

Make sure part #28 is pushed up against this shoulder

Part#28

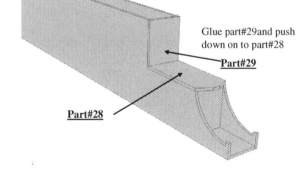

Glue step#7
Glue part#29 on to the assembly

Glue part#29 and push down on to part#28

Part#29

Part#28

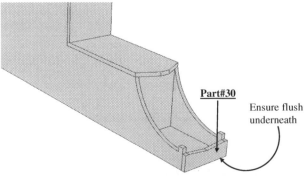

Glue step#8
Glue one piece of part#30 to the assembly

Part#30

Ensure flush underneath

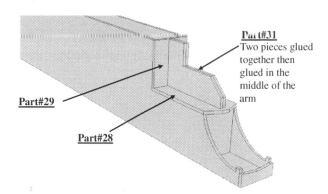

Par t#31
Two pieces glued together then glued in the middle of the arm

Part#29

Part#28

Glue step#9
Glue two pieces of part#31 together then glue to assembly. Make sure the two pieces are centred on part#28 and part#29

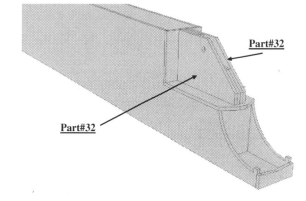

Part#32

Part#32

Glue step#10
Glue two pieces of part#32 either side of the two pieces of part#31. This arrangement will form a hinge system for the finger

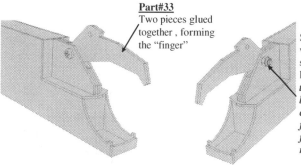

Part#33
Two pieces glued together , forming the "finger"

Secure the finger with 10 mm long screw and a locknut. *Do not tighten the nut because you will clamp the finger. The finger should rotate freely*

Glue step#11
Glue two pieces of part#33 together, wait a few minutes for the glue to dry then enter carefully into the hinge system and secure with an M3 stainless steel pan crosshead screw, 10mm long.

Part#34
Two pieces glued together

44 mm

Glue the two pieces 44 mm from end

Glue step#12
Glue two pieces of part#34 on to the arm positioned as shown. These two stacked pieces restrict the elastic band which will snap shut the finger.

Now check and examine that the ping-pong ball
fits into the three-point mounting system ⊡ ⊡⊡
and that the finger holds the ball with a fourth
point which is the finger tip

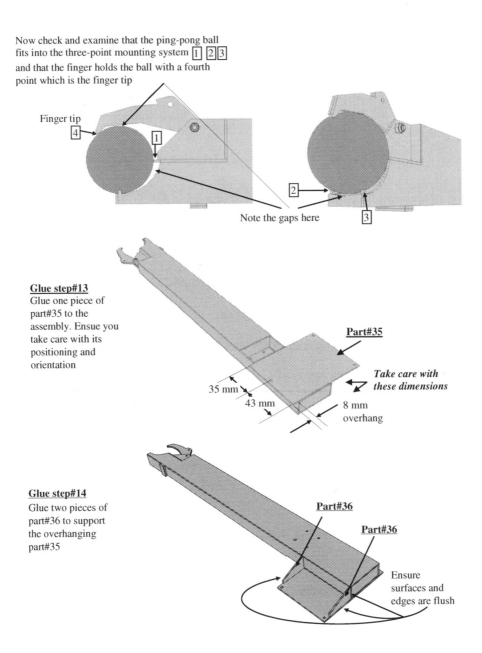

Finger tip
⊡

⊡

⊡

⊡

Note the gaps here   ⊡

Glue step#13
Glue one piece of
part#35 to the
assembly. Ensue you
take care with its
positioning and
orientation

Part#35

Take care with
these dimensions

35 mm

43 mm

8 mm
overhang

Glue step#14
Glue two pieces of
part#36 to support
the overhanging
part#35

Part#36

Part#36

Ensure
surfaces and
edges are flush

Now screw on three pieces of 15 mm x M3 plastic stand-off pillars. Use M3 x 6 mm long plastic screws with pan crosshead.

Here are the screws as viewed from underneath

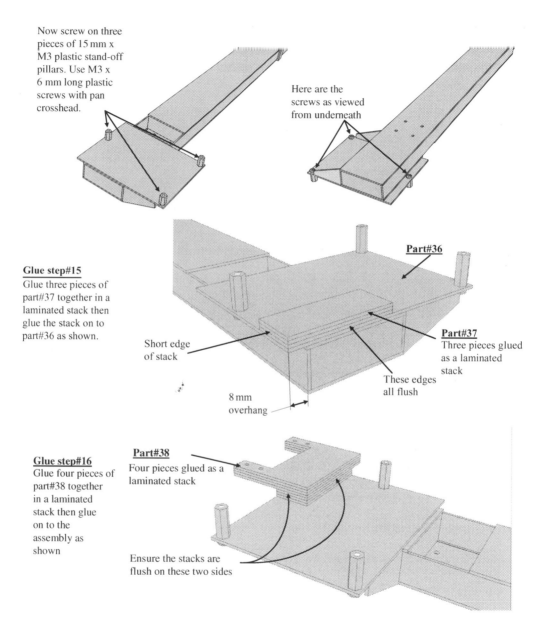

Glue step#15
Glue three pieces of part#37 together in a laminated stack then glue the stack on to part#36 as shown.

Part#36

Short edge of stack

Part#37
Three pieces glued as a laminated stack

These edges all flush

8 mm overhang

Glue step#16
Glue four pieces of part#38 together in a laminated stack then glue on to the assembly as shown

Part#38
Four pieces glued as a laminated stack

Ensure the stacks are flush on these two sides

Screw a Futaba S9452 or BLS471SV or equivalent servo on to the part#38 stack. Use four stainless steel screws, M3 pan crosshead x 12mm long. Use the rubber gaiters supplied with the servo. Set the mid range position of the horn to the position shown here in the diagram, see Chapter 7 Datuming the Horn, Fig. 7.7. Use a horn that has a hole set at 19 mm ± 2 mm. Drill hole to ϕ3.0 mm

Drill hole ϕ3.0 mm

90° horn at mid-range position

Servo secured with 4 screws self-tapped into the ϕ2.5 mm drilled holes in the part#38 stack

19 mm ± 2 mm horn radius

Part#38 stack

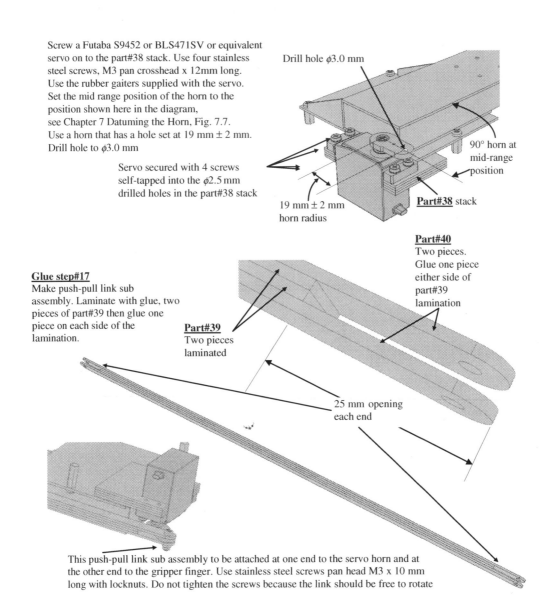

Part#40
Two pieces.
Glue one piece
either side of
part#39
lamination

Glue step#17
Make push-pull link sub assembly. Laminate with glue, two pieces of part#39 then glue one piece on each side of the lamination.

Part#39
Two pieces
laminated

25 mm opening
each end

This push-pull link sub assembly to be attached at one end to the servo horn and at the other end to the gripper finger. Use stainless steel screws pan head M3 x 10 mm long with locknuts. Do not tighten the screws because the link should be free to rotate

Figures showing smart arm push-pull link in three positions which are:
  (i) Fully open position
 (ii) Mid-range position
(iii) Ball clamped position

By programming the smart arm Basic Stamp microcomputer you should check these opening, mid-range, and closing positions.

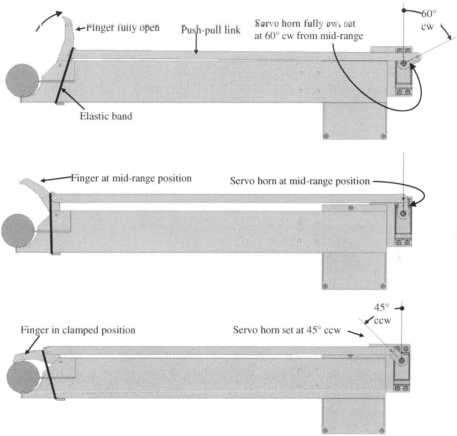

Finger fully open ← Push-pull link   Servo horn fully cw, set at 60° cw from mid-range   60° cw

Elastic band

Finger at mid-range position   Servo horn at mid-range position

Finger in clamped position   Servo horn set at 45° ccw   45° ccw

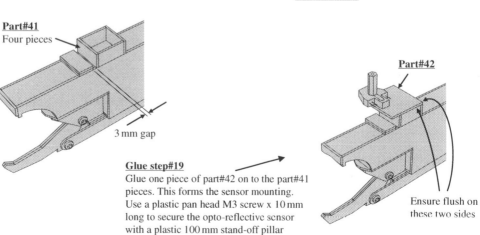

Glue step#18
Glue four pieces of part#41. Look closely to see the edge overlapping arrangement of fitting the pieces together. The 3 mm gap is to confine the elastic band

Part#41
Four pieces

3 mm gap

Part#42

Glue step#19
Glue one piece of part#42 on to the part#41 pieces. This forms the sensor mounting. Use a plastic pan head M3 screw x 10 mm long to secure the opto-reflective sensor with a plastic 100 mm stand-off pillar

Ensure flush on these two sides

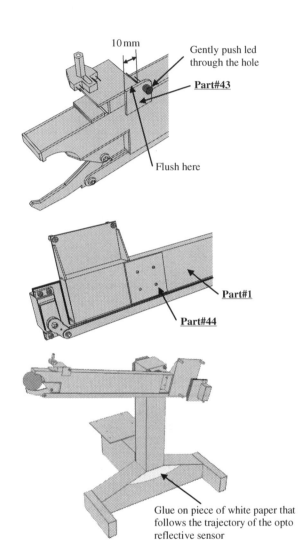

Glue step#20
Glue one piece of part#43 on to the sensor mounting. This is the mounting for an led that is used for indicating ball release.

10 mm

Gently push led through the hole

Part#43

Flush here

Glue step#21
Glue one piece of part#44 on to the back of the smart arm making sure that the four holes align with the four holes on part#1

Part#1

Part#44

Glue step#22
Glue one piece of white paper on the structure as pictured such that the reflective opto sensor can detect a reflective surface. Be careful that black cardboard can also be sensed as a bright surface. You need to do tests to find out. The plastic stand-off pillar that secures the opto reflective sensor should be adjusted so that the sensor is 5 mm away from the white paper, i.e., a 5 mm gap. This gives the best reflective signal.

Glue on piece of white paper that follows the trajectory of the opto reflective sensor

This concludes the mechanical construction of the Throwing robot. Now we move to its electrical wiring

10.3 ELECTRICAL WIRING OF THE STEPPER MOTOR

There are two separate and noninteracting circuits and microcomputers systems for the Throwing robot. The first system is the stepper motor system that controls the rotation of the smart arm. The arm has to be put in a vertical position by hand before the arm is rotated. This is the datum starting position. The software code then accelerates the arm to maximum angular velocity in one revolution where it maintains this maximum speed for one more revolution during which time the smart arm releases the ball then the arm slows down to a stop in one more revolution. Thus the software code drives the arm through three revolutions.

The circuit diagram for the stepper motor Basic Stamp BS2px that controls the Leadshine DM542 stepper motor driver is shown in Fig. 10.1. The BS2px microcomputer is about 2.5 times faster at implementing instructions than the BS2 so it can drive the smart arm at a higher speed than the BS2.

The photograph in Fig. 10.2 shows that how the arm rotation system is wired up.

10.4 PROGRAMMING THE STEPPER MOTOR TO ROTATE THE SMART ARM

The software code for driving the stepper motor is now given.

The BS2px programme for the arm rotation stepper motor

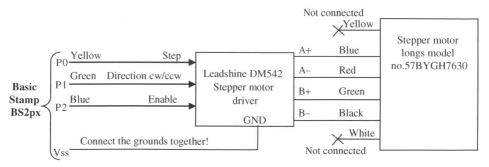

■ **FIG. 10.1** Circuit diagram of the smart arm stepper motor rotation drive system.

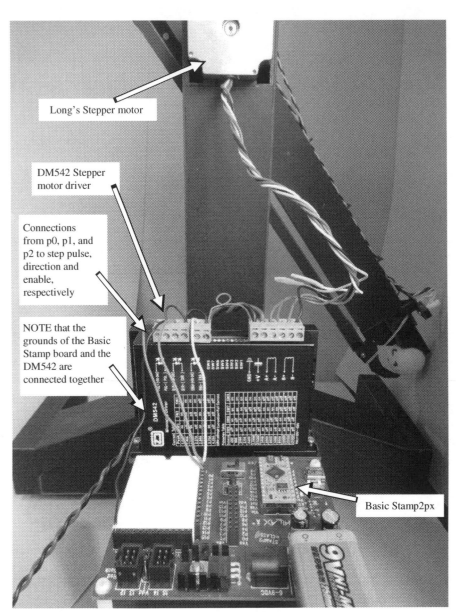

Long's Stepper motor

DM542 Stepper motor driver

Connections from p0, p1, and p2 to step pulse, direction and enable, respectively

NOTE that the grounds of the Basic Stamp board and the DM542 are connected together

Basic Stamp2px

■ **FIG. 10.2** Photograph showing the wiring of the arm rotation stepper motor system.

```
'The DM542 driver switch settings are set as follows:
'SW1=OFF, SW2=ON, SW3=OFF, SW4=ON, SW5=OFF, SW6=ON, SW7=ON, SW8=ON
'This means that the phase current is set to 3.03A and half step mode 400step/rev is activated
'Two 12V power supplies are used that are connected in series to give 24V supply.
'The coils are rated at 6V but the DM542 driver is a chopper driver that restricts the maximum
current.
'Maximum current here is restricted to 3.03A.
'The programme accelerates the arm from its pull-in speed in 400 steps (1 revolution) to its
maximum speed
'The programme maintains this maximum speed for 400 more steps, i.e. 1 revolution
'The programme then decelerates in 400 steps, i.e. 1 revolution, back down to its pull-in speed
then stops.
enable      CON 0         'pin0 is the enable pin, low=enable driver, high=disable driver
dir         CON 1         'pin1 is the direction pin, high=ccw, low=cw (throwing direction)
step        CON 2         'pin2 is the step pin, 2µs positive pulse makes the stepper motor move
                          by 'one step
a           VAR Word      'variable used for the stepping period
i           VAR Word      ' general variable used for counting
LOW enable  'enable motor driver
            LOW 1         'enable throwing direction
'again1:    PAUSE 1000    ' wait 1 second before starting programme
            a=4200        'set pull-in period to 4200 x 0.81µs=3.40ms/step =3.40ms/step x
       '400step/rev=1.36s/rev. Thus starting speed=0.735rev/s
       FOR i=1 TO 400     'accelerate in 400steps, i.e. 1 revolution
       PULSOUT step,3     'rotate by 1 step
       PULSOUT 3,a        'delay for period
       a=a-9              ' next step period is 9 x 0.81µs less so accelerating
       'final period=(4200-(400 x 9)) x 0.81µ Thus approx 5rev/s but there will be
       'additional periodic delay due to the For-Next loop
       NEXT
       FOR i=1 TO 400 'stay at maximum speed for 400steps, i.e. 1 revolution to give opportunity
       'for the smart arm to let go the ball
       PULSOUT step,3     'one step rotate
       PULSOUT 3,a        'periodic delay is constant
       NEXT
       FOR i=1 TO 400     ''decelerate in 400steps, i.e. 1 revolution
       PULSOUT step,3     'one step rotate
       PULSOUT 3,a        'periodic delay
       a=a+9              'increase the periodic delay in order to decelerate
       NEXT
       PAUSE 1000         'delay 1 sec to give time for the arm to stop vibrating
       HIGH enable        ' 'disable motor driver
again: GOTO again         'infinite loop to stop programme, end of stepper motor programme
```

10.5 **The Smart Arm Electrical Circuitry**

Fig. 10.3 shows the second circuit diagram and microcomputer for the smart arm control. The system is mounted on the smart arm complete with a 9 V rechargeable NiMH battery. The circuit diagram shows the Basic Stamp2 pin connections to the optoreflective sensor, the indicator led light and the servo. The optoreflective sensor has an inbuilt infrared transmitter (Tx) and receiver (Rx). The transmitter is set at angle such that when a surface is 5 mm away the maximum light will be reflected into the receiver. The receiver, Rx conductance, on receiving light intensity will increase, in other words, the resistance will decrease and thus the voltage measured by Basic Stamp pin, p6 will drop. The $10 k\Omega$ pull-up resistance is a value such that when light is reflected off a white paper surface, the voltage at p6 gives a LOW state, meaning less than 0.25 V. Thus, white paper is indicated by reading a zero with the IN6 instruction.

The photograph, Fig. 10.4, shows the wiring according to Fig. 10.3 circuit diagram (Fig. 10.5).

Now we move on to showing a sample program for the smart arm.

10.6 **REAL-TIME PROGRAMMING THE SMART ARM**

Remember the Throwing robot has two separate noninteracting programs; one program for the stepper motor driver which is the Leadshine DM542.

```
' {$STAMP BS2px}
Programme for the Smart arm
' {$STAMP BS2}
'program for smart arm
'This programme uses an edge-triggered software method, in
conjunction with a reflective opto-sensor, to 'measure the arm
```

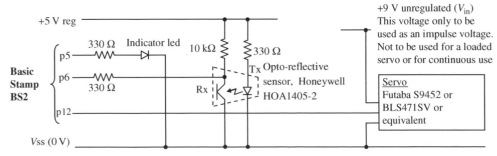

■ **FIG. 10.3** Circuit diagram of the smart arm.

To the smart arm finger gripper end

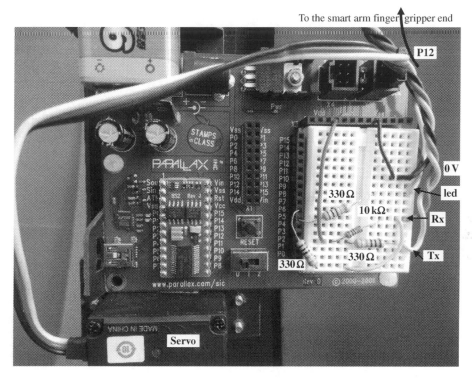

■ **FIG. 10.4** Photo of smart arm circuitry.

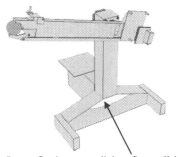

Opto reflective sensor light reflects off this white paper but be careful because the infrared light also bounces off black cardboard in near-equal measure. That's why the paper covers the whole of the path of the light sweep across the cardboard tower and base

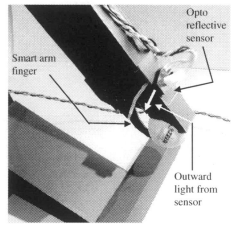

■ **FIG. 10.5** Photograph of the finger end of the smart arm showing the optoreflective sensor. Picture on left hand side shows the white paper reflective surface.

passing a white surface. Following two passes of the white surface, the smart arm software 'starts a clock that waits a given amount of time to allow the arm to reach a 40° launching angle then to activate 'the servo that releases finger to let go and throw the ball. The edge-triggered software simulates a finite state, '2-state logic machine that could be implemented with digital gates but here it is implemented in software. The 'disadvantage of the software implementation is its lack of speed which will lead to repeatability errors in the 'launch angle. As such a valuable and interesting piece of research can be done in evaluating this repeatability 'error. Experiments need to be done to determine the servo PULSOUT values for the finger in its open and 'closed 'positions. Remember that you don't have to power on the servo to keep it open or closed. The elastic 'band and its positioning either side of the finger joint means that it is a mechanical latch that when the servo is 'impulse open or closed for a short time then it can be turned powered off and the finger will stay open or closed 'respectively.

'PULSOUT values for finger are, 550=finger fully open, 950=finger fully closed
'pin 5 is the led
'pin 6 is the reflective opto sensor, low = white surface, i.e. "white", high = no reflection, i.e. "black"
'pin12 is the servo

```
led          CON   5      'Pin 5 is the led indicator control pin
i            VAR   Word   'variable used for counting
j            VAR   Nib    'used as reflective opto sensor trigger count
             j=0
             LOW led      'make sure led indicator is switched off
```

'Software edge trigger algorithm finite state machine changes state as arm passes from "black" to "white"

```
state1:      IF IN6=0 THEN state2    'sensor is "white" so change to state2
             GOTO state1             'stay at state1 but check if should change to state2
state2:      j=j+1                   'increase arm revolution count number
             IF j>1 THEN fingeropen  'open the finger on second revolution
skip1:       IF IN6=1 THEN state1    'sensor is black so go reset to state1
             GOTO skip1              'state is still state1 but check if should change state
fingeropen:  PAUSE 90                ' open the finger after 90ms to give required launch angle
             FOR i=1 TO 20           'open the finger by impulsing the servo for 120ms approx
             PULSOUT 12, 550
             PAUSE 5
             NEXT
             HIGH led                'turn on the led to indicate servo opened
             PAUSE 1000              'wait for 1 second
             LOW 5                   'turn off the led
again1:      GOTO again1             'infinite loop to stop programme
```

That concludes the mechanical and electrical construction and programming of the Throwing robot. Some problems are discussed in the following section.

10.7 **PROBLEMS**

1. In the game of cricket you can bowl underarm or overarm. Underarm is never used in cricket because it is less effective at delivering a ball at high speed. Program the robot to bowl underarm and examine if it is just as effective or not as overarm bowling, for example, does underarm throwing give better ball-in-basket accuracy.

2. Is it possible to throw a ball for a distance of 6 m with only one revolution of the smart arm? In other words how can you accelerate to 12 m/s launch speed in one revolution?

3. Design and implement and experiment that determines the repeatability error of the launch angle.

4. Evaluate the increase in accuracy, that is, the decrease in repeatability error, if a higher speed BS2px microcomputer is used for the smart arm.

5. Design a sensor system for the stepper motor microcomputer that datums the angular position of the smart arm. The system obviates the need for manually datuming the arm to the upright position before throwing.

6. Design and build a small lightweight wound coil and magnet sensor or Hall-effect sensor that the smart arm uses to datum itself. This should make the machine more accurate at releasing the ball at the precise time. Better still is to check automotive solutions such as those produced by Allegro speed sensing systems. Automotive sensor and actuator systems are amazing because they are cheap and highly reliable.

Theory VII: Theory and Design Notes of the Catapult Robot

LEARNING OUTCOMES

1. Understanding of how to create a precision single-axis slideway.
2. Analysis of a nonlinear coiled spring system to store and release mechanical energy.

11.1 OVERVIEW OF THE CATAPULT LAUNCHING SYSTEM

Fig. 11.1 illustrates the catapult arrangement which is composed of a twin aluminum parallel tube structure that forms a single-axis slideway for a shuttle that houses a ping-pong ball that is to be catapulted out of the shuttle by a spring system. The aluminum tubes act as low-friction slideway runners for the shuttle. Aluminum circular tubes are available from hardware stores at low cost. The tubes used in the catapult robot have outside and inside diameters of 16 and 14 mm, respectively. They have high precision tolerances in terms of roundness, smoothness, straightness, and diameter and have the bonus of being lightweight, stiff, and strong.

The spring system is composed of four individual springs mounted symmetrically above and below the centerline and to the left and right of the centerline.

11.2 THE SPRING ENERGY STORAGE CATAPULT SYSTEM

A pull-back force displaces the shuttle against a spring force system composed of four individual springs. The pull-back force is generated from a servo that is utilized in Chapter 12 "Catapult Construction." The shuttle is a special design with four registration shapes at each of its corners. These shapes permit the shuttle to slide precisely along the tube runners with low friction whereas motion in the remaining five degrees of freedom of the shuttle is inhibited.

Creating Precision Robots. https://doi.org/10.1016/B978-0-12-815758-9.00011-0

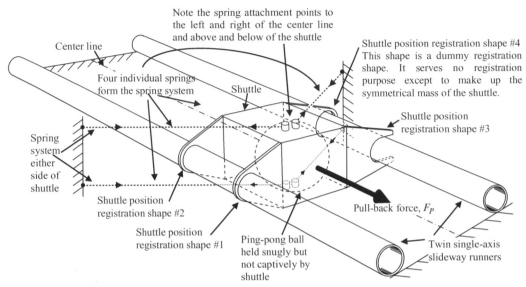

Note the spring attachment points to the left and right of the center line and above and below of the shuttle

Center line

Four individual springs form the spring system

Shuttle

Shuttle position registration shape #4 This shape is a dummy registration shape. It serves no registration purpose except to make up the symmetrical mass of the shuttle.

Shuttle position registration shape #3

Spring system either side of shuttle

Shuttle position registration shape #2

Shuttle position registration shape #1

Ping-pong ball held snugly but not captively by shuttle

Pull-back force, F_P

Twin single-axis slideway runners

■ **FIG. 11.1** The shuttle in its single-axis slideway. Note the four shuttle position registration shapes #1, #2, #3, and #4 which are fundamental for the precise, low friction, single-axis motion of the shuttle.

Fig. 11.2 shows a plan view of the spring energy storage system which is composed of four steel-coiled springs mounted symmetrically either side of a centerline that catapults the ping-pong ball up to a distance of 6m. The springs can be replaced with elastic bands. Such a replacement represents an interesting exercise for students to choose how many elastic bands and whether to put them in series or parallel to give the correct elasticity. There is an identical mirror-image spring system, in the plane of the paper, immediately beneath the spring system. This can be seen by referring back to Fig. 11.1. Thus there is a total of four individual springs arranged as two pairs of individual springs.

So the spring system is symmetrical both about the plane of the paper and about the center line. The reason for this is that (i) the springs are able to apply a catapulting force vector, F_P whose line of action passes through the center of the ball and (ii) that there are no force components from F_P that are in or out of the paper or to the left or right of the centerline. Thus the spring system will reduce to near zero any friction forces between the slideway runners and the shuttle. To satisfy these requirements, the spring attachment points to the shuttle are attached symmetrically above and below the paper plane and to the left and right of the centerline, Fig. 11.1.

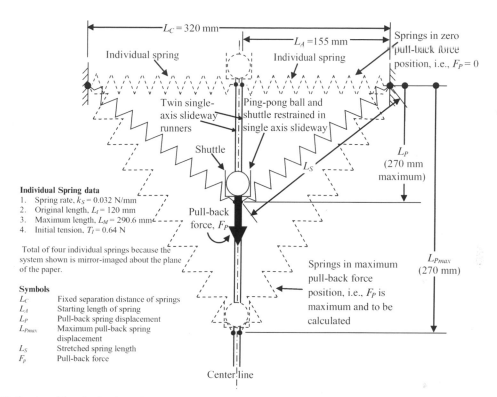

FIG. 11.2 Plan view of the catapult spring system.

A mathematical analysis of the stored energy and pull-back force of the spring system will be done shortly. Before we do this, it is relevant to talk more about the design of the shuttle and slideway systems.

11.3 **THE SHUTTLE DESIGN**

The sliding shuttle, Fig. 11.3, has a noncaptive receptacle for housing the ping-pong ball plus four registration shapes, numbered, #1, #2, #3, and #4 that protrude from its corners. These shapes permit the low-friction precision sliding of the shuttle along parallel aluminum tube runners. Precision low-friction sliding of the shuttle is fundamental to the accurate launching velocity of the ping-pong ball and is one example of why this book is entitled "Creating Precision Robots."

Fig. 11.4 shows registration shapes#1 and #3. Shape#1 inhibits shuttle motion in the *x*- and *y*-axis displacements but permits low-friction sliding in the *z* axis which is along the slideway runners. Shape#3 inhibits rotation

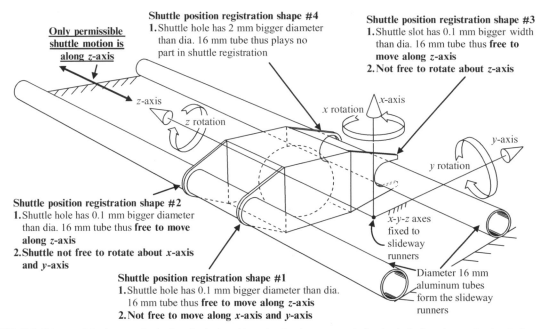

Only permissible shuttle motion is along z-axis

Shuttle position registration shape #4
1. Shuttle hole has 2 mm bigger diameter than dia. 16 mm tube thus plays no part in shuttle registration

Shuttle position registration shape #3
1. Shuttle slot has 0.1 mm bigger width than dia. 16 mm tube thus **free to move along z-axis**
2. **Not free to rotate about z-axis**

z-axis

x rotation

z rotation

x-axis

y-axis

y rotation

Shuttle position registration shape #2
1. Shuttle hole has 0.1 mm bigger diameter than dia. 16 mm tube thus **free to move along z-axis**
2. **Shuttle not free to rotate about x-axis and y-axis**

x-y-z axes fixed to slideway runners

Shuttle position registration shape #1
1. Shuttle hole has 0.1 mm bigger diameter than dia. 16 mm tube thus **free to move along z-axis**
2. **Not free to move along x-axis and y-axis**

Diameter 16 mm aluminum tubes form the slideway runners

■ **FIG. 11.3** Slideway and shuttle system showing how the shuttle position registration shapes create a single axis only shuttle motion meaning that shuttle motion is only permissible along the z-axis. Motion along x-axis and y-axis is not possible and neither is rotation about x, y, and z axes. Thus of the six possible degrees of freedom of the shuttle, five are inhibited thus leaving only one in the z-axis direction that is permissible.

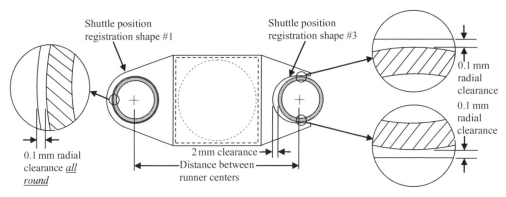

Shuttle position registration shape #1

Shuttle position registration shape #3

0.1 mm radial clearance

0.1 mm radial clearance

0.1 mm radial clearance *all round*

2 mm clearance
Distance between runner centers

■ **FIG. 11.4** View showing the shuttle position registration shapes#1 and #3. The aluminum tubes which serve as the slideway runners are shown as *shaded concentric circles.*

about the z-axis but is tolerant to the distance variation between runner centers that is related to lack of parallelness of the runners, Fig. 11.5.

Note in Fig. 11.4 (i) the 0.1 mm clearances between the tube runner and shapes#1 and 3 and (ii) the 2 mm clearance on the inside of shape#3.

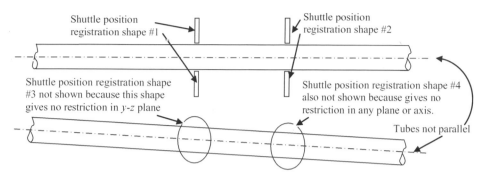

FIG. 11.5 Plan view of runners showing tube *parallel* alignment error and how the shuttle is tolerant to these errors.

Fig. 11.6 shows registration shape#2 that inhibits shuttle rotation about x and y axes. Thus shapes#1, #2 and #3 inhibit the shuttle in five degrees of freedom and permit only motion along the z-axis. Only three of the four registration shapes serve a registration purpose. The fourth shape, registration shape#4, is a "dummy" shape and serves only the purpose of keeping the center of mass on the centerline of the shuttle that is coincident with the center of the spring system force.

Thus the spring system force vector that accelerates the shuttle and the d'Alembert dynamic inertial reaction shuttle force vector lie mutually coaxial. This means that there will be no couples that need to be created for dynamic equilibrium by the slideway runners. These couples would produce frictional retarding forces leading to low efficiency and low precision of launching speed.

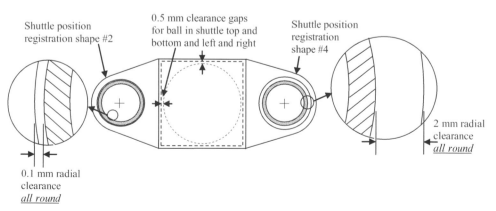

FIG. 11.6 View showing the shuttle position registration shapes#2 and #4. Once again, the aluminum tubes which serve as the slideway runners are shown as *shaded concentric circles*. Note also the ball/shuttle clearance gaps that permit runner tolerances.

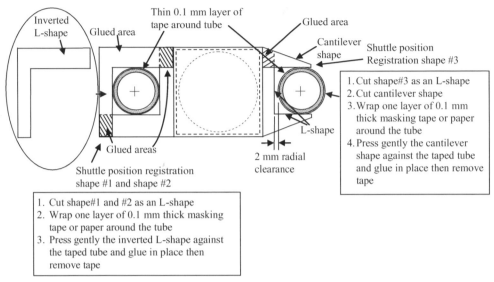

■ **FIG. 11.7** Practical method of obtaining very accurate 0.1 mm radial clearance tolerances.

Fig. 11.7 shows a practical method and instructions for obtaining the 0.1 mm tolerances for the registration shapes. Fig. 11.8 shows the permissible tube error tolerance angle of 4 degrees due to the 0.1 mm radial clearance in registration shapes #1 and #2. This tube angle tolerance angle leads to a maximum permissible coplanar error when viewed from the side, Fig. 11.9. Tube angle error also leads to a twisting of the shuttle as it moves along the slideways. However, it is not difficult to achieve a tube angle tolerance angle of well less that 4 degrees; less than 0.2 degrees is possible.

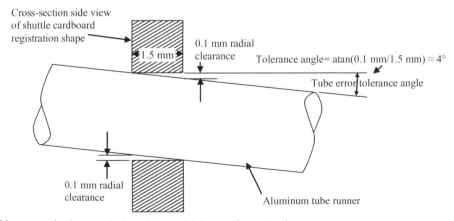

■ **FIG. 11.8** Tube runner angle tolerance angle due to 0.1 mm radial clearance of registration shapes.

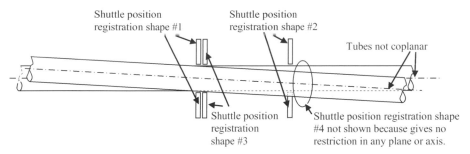

■ **FIG. 11.9** Side view of runners showing tube *planar* alignment error and how the shuttle is tolerant to this error. Note that if the runners are not coplanar then the shuttle will experience a *z*-axis rotation as it moves along the runners in the *z* direction.

We now conclude the description of the shuttle design and move on to the spring system mathematical analysis.

11.4 MATHEMATICAL ANALYSIS OF THE CATAPULT SPRING SYSTEM

Fig. 11.10 shows the shuttle pulled back leftwards from a starting shuttle position#1 where pull-back displacement, $L_P = 0$ (mm) and pull-back force, $F_P = 0$ (N) to a shuttle position#2 where the displacement, L_P has been produced by a force F_P. Here the shuttle stays static until it is released. The shuttle then accelerates toward the right direction to a maximum speed, assuming zero friction, at the starting shuttle position. After passing through the starting shuttle position, the spring force acts in the opposite direction

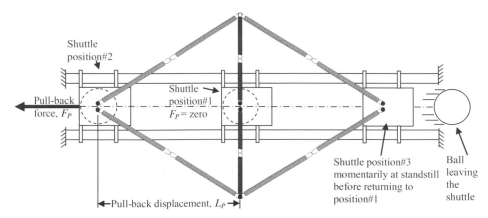

■ **FIG. 11.10** Plan view of catapult spring system in three positions. *Position#1* is starting position before pull-back force applied; *position#2* is at pull-back position before release of the shuttle and *position#3* is the shuttle at zero velocity after the ball has left the shuttle and before the shuttle returns to position#1. Note that position#1 also is the position where the shuttle is at maximum speed moving to the right before it starts to decelerate due to the spring pulling the shuttle to the left.

toward the left and decelerates the shuttle thus decreasing its speed where-upon the ball, unrestrained by the shuttle, carries on toward the right at the maximum speed and leaves the shuttle which is slowing down. Eventually the shuttle comes to a standstill momentarily at shuttle position#3 then restores its position, after oscillating, at shuttle position#1.

We now analyze the spring force as the shuttle is pulled back. The idea is to obtain the pull-back force as a function of the displacement of the shuttle. From this relationship we can obtain the strain energy in the springs as a function of the displacement of the shuttle. When the shuttle is let go, the spring strain energy is transferred to the shuttle and the ping-pong ball as kinetic energy thus, producing an accurate launching speed necessary to cat-apult the ping-pong ball into a target basket. The shuttle needs to be as low mass as possible otherwise an excessive amount of energy is lost in giving the kinetic energy to the shuttle when the whole idea is to impart kinetic energy to the ping-pong ball, not the shuttle.

We use Misumi springs and one "individual spring" is made up from two Misumi, AUA 6-60 springs in series, Fig. 11.11.

Referring to Fig. 11.1, the pull-back force, F_P is given by the following reasoning:

$$\text{First of all, individual spring force,} F_S = (L_S - L_I) \times k_S + T_I \qquad (11.1)$$

…where,

L_S is the individual stretched spring length
L_I is the initial or starting spring length
k_S is the individual spring rate constant
T_I is the initial spring tension, that is, the force required to initiate extension

$$\text{Now, the spring stretched length,} L_S = \sqrt{(L_A{}^2 + L_P{}^2)} \qquad (11.2)$$

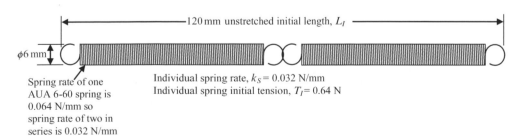

120 mm unstretched initial length, L_I

ϕ6 mm

Spring rate of one AUA 6-60 spring is 0.064 N/mm so spring rate of two in series is 0.032 N/mm

Individual spring rate, $k_S = 0.032$ N/mm
Individual spring initial tension, $T_I = 0.64$ N

■ **FIG. 11.11** One "individual spring" composed of two Misumi AUA 6-60 stainless steel springs hooked together in series.

We need to calculate the pull-back force, F_P that is the product of four individual springs,

$$..\text{thus, } F_P = 4 \times F_S \times \frac{L_P}{L_S}$$

Four individual

springs

(11.3)

Now substitute Eqs. (11.1), (11.2) into Eq. (11.3) to obtain, pull-back force, F_P as follows:

$$\text{Pull-back force, } F_P = 4 \times \left\{ \sqrt{(L_A{}^2 + L_P{}^2)} - L_I \right\} \times k_S + T_I \} \times \frac{L_P}{\sqrt{(L_P{}^2 + L_A{}^2)}}$$

(11.4)

Eq. (11.4) now needs to be integrated with respect to L_P in order to obtain the work done on the spring. This work done represents the strain energy stored in the spring system assuming there is no hysteresis in the springs. From the strain energy we can obtain the kinetic energy transferred to the ball and shuttle and hence the launching speed of the ball. However, Eq. (11.4) is neither nice nor convenient to integrate so a polynomial fit equation will be used to approximate accurately the relationship of Eq. (11.4).

Substituting the following values into Eq. (11.4),

$L_A = 0.155\,\text{m}$
$L_I - 0.120\,\text{m}$
$k_S = 32\,\text{N/m}$
$T_I = 0.64\,\text{N}$

…we get the relationship shown in Fig. 11.12 for pull-back force, F_P as a function of pull-back displacement, L_P. This relationship is accurately approximated to within 0.5% of the maximum pull-back force with the following polynomial equation:

$$\text{Pull-back force, } F_P = 36.4L_P + 251L_P{}^2 - 232L_P{}^3$$

(11.5)

…where L_P is the pull-back displacement.

It is now straightforward to integrate Eq. (11.5) with respect to L_P in order to obtain the strain energy stored in the spring system.

Assuming no losses during spring extension, the strain energy, E_S stored by the spring system is given by:

$$E_S = \text{work done on spring} = \int F_P \cdot dL_P$$

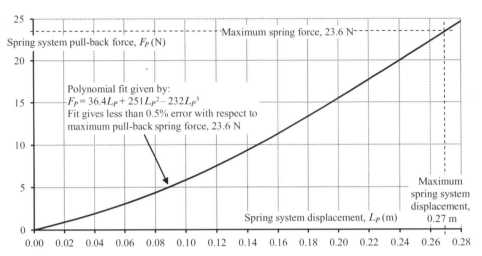

■ FIG. 11.12 Graph showing nonlinearity of spring system pull-back force versus displacement. Note the polynomial fit equation of the relationship. The polynomial equation enables its convenient integration with respect to displacement to give the spring system stored strain energy.

$$\text{Thus, } E_S = \int \left(36.4L_P + 251L_P^2 - 232L_P^3 \right) \cdot dL_P$$

Hence,

$$\text{Spring strain energy, } E_S = 18.2L_P^2 + 83.7L_P^3 - 58.0L_P^4 \qquad (11.6)$$

We now make, spring strain energy, E_S equal to the kinetic energy of the shuttle mass plus ping-pong ball mass, m_{SB} and solve for the pull-back displacement, L_P, given the launching speed, v_L as follows:

Assuming no losses during ball launching and that the slideway is horizontal so no attention paid to the potential energy loss or gain of the shuttle plus ball:

$$\text{Spring strain energy, } E_S = \text{kinetic energy of shuttle plus ball} = \frac{1}{2} \times m_{SB} \times v_L^2$$

$$\text{Thus}: E_S = 18.2L_P^2 + 83.7L_P^3 - 58.0L_P^4 = \frac{1}{2} \times m_{SB} \times v_L^2$$

$$\text{Hence, launching speed, } v_L = \text{sqrt} \left\{ \frac{2 \times \left(18.2L_P^2 + 83.7L_P^3 - 58.0L_P^4 \right)}{m_{SB}} \right\}$$

$$(11.7)$$

Eq. (11.7) is plotted calculated with Excel spreadsheet then plotted, not v_L as a function of L_P, but inverted with L_P as a function of v_L, Fig. 11.13. Then an accurate polynomial fit, accurate to within 0.3% of maximum pull-back displacement, is made to the relationship, given by

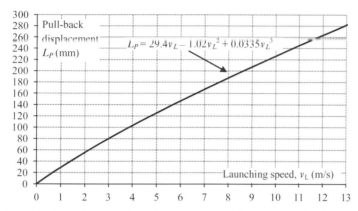

$$L_P = 29.4v_L - 1.02v_L^2 + 0.0335v_L^3$$

■ **FIG. 11.13** Relationship that can be used to compute pull-back displacement given launching speed.

Pull-back displacement, $L_P = 29.4v_L - 1.02v_L^2 + 0.0335v_L^3$ (11.8)

...where, v_L is the launch speed.

11.5 **LAUNCHING SPEED PRECISION**

Thus given a required launching speed, v_L which has been computed from the range of the target, Eq. (11.8) can be used to compute the pull-back displacement, L_P to produce the required launching speed. The chapter concerned with constructing the catapult robot will show how the pull-back displacement is achieved with a servo and pulley system. Let us now demonstrate how Eq. (11.8) is used. First of all the range of the target has been measured by, for example, a computer vision system or simply by a tape measure. Next the theory of trajectory computation in Chapter 3 is used to compute a value of required launching speed and launching angle. The chapter on constructing the catapult robot will be concerned with setting the launch angle. Here we are tasked with computing the pull-back displacement, L_P in order to obtain the correct launching speed. Suppose that a 10.0 m/s launching speed is required. Note that we want 10.0 m/s, meaning a speed between 9.95 and 10.05 m/s, that is, 10.0 ± 0.05 m/s. This will give a ball catapulting range of 5.16 ± 0.03 m which has been considered a required range accuracy, let us say, by a student allocated to computing the precision of the launching velocity. Now, substituting $v_L = 10.0 \pm 0.05$ m/s into Eq. (11.8) we obtain a pull-back displacement, $L_P = 225.5 \pm 1.0$ mm hence the precision of pull-back displacement must be within ± 1 mm tolerance. The range error due to the launch angle has to be taken into account also but we are only concerned here with launching speed precision. By the time students have built and tested their catapult robot, they will be becoming acutely aware of precision and accuracy, and its importance, in engineering design.

The slideway and spring system has been designed to enable the safe dissipation of the shuttle kinetic energy after the ball has left the shuttle. The method is simply to let the shuttle overshoot up to a maximum of 270 mm which is the maximum pull-back displacement, L_{Pmax}, Fig. 11.2. Eventually the low-friction slideway cum shuttle will dissipate the energy and the shuttle will come to rest at its mid position. Fig. 11.14 indicates approximately what happens. Note that the if the pull-back displacement is the maximum of 270 mm then the ball will be launched at approximately 12 m/s and leave the shuttle at a shuttle displacement, $L_P = 0$ mm.

As a final note, students should recall Newton's third law which means that as the stretched springs are released and the shuttle cum ball is accelerated forward by the spring force, there will be an equal and opposite reactionary force backward. This force impulse will jerk the catapult structure backward

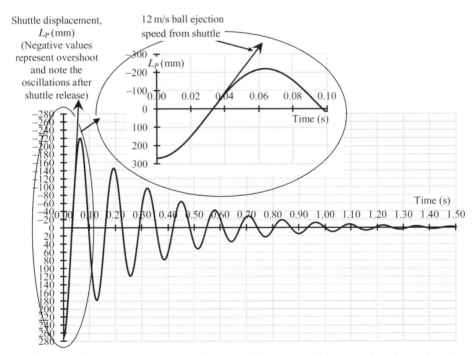

■ **FIG. 11.14** Approximate shuttle displacement as a function of time after its release. Note that positive displacement is shown downwards since it represents pull back spring displacement that is pulled in the downward direction. Thus upward shuttle displacement is shown as negative values and is an oscillatory overshoot after the ball leaves the shuttle. The time decaying amplitude of oscillations represents the eventual dissipation of shuttle kinetic energy largely dissipated by raising the temperature of the slideway due to friction between shuttle and slideway.

and dissipate some of the much needed forward force impulse thus lowering the launching speed of the ball. It is akin to appreciating that the ball is being "hit," not propelled, by the shuttle where the coefficient of restitution (see collision theory, Chapter 5) between the ball and the shuttle is less than unity and hence there is an energy loss that is dissipated by the catapult structure.

11.6 **PROBLEMS**

1. Examine how much the launching angle will affect the ball launching speed. For example the potential energy gain of the ball and shuttle is a function of the angle of launch. In other words, the launching speed at a horizontal launching angle (0 degree) will be greater than the launching speed at a vertical launching angle of 90 degrees.

2. Investigate the implications and design challenges of a reaction-less launching system. In other words, there is no jerk as the ball is launched and as much as possible of the spring stored energy goes into the ball and not into jerking the catapult main structure backwards with an associated loss of ball momentum.

3. Investigate a launching system that uses the full length of the slideway to accelerate the ball. The shuttle is decelerated to a standstill in a short distance with a shock absorber. You can check out Misumi shock absorbers. What advantage, if any, is there in having a longer acceleration distance?

4. Confirm, or not, the pull-back spring displacement versus pull-back force relationship calculation by measuring the values.

The Catapult Robot: Design and Construction

LEARNING OUTCOMES

1. Understanding how to make a precision efficient single-axis slideway.
2. How to make a servo-actuated cocking and trigger system.
3. Increased knowledge and experience of designing, building, and programming a mechatronic system.

The drawings and building instructions will be presented in stages. We start with the base structure then follow with the single-axis slideway and finally the electrical wiring and programming.

12.1 BASE STRUCTURE DRAWINGS

All dimensions in millimeter and material is 1.5 mm cardboard.

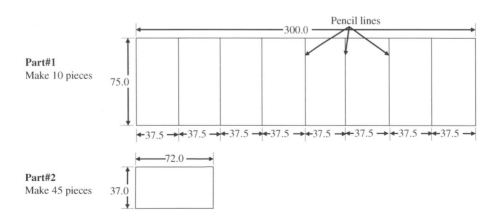

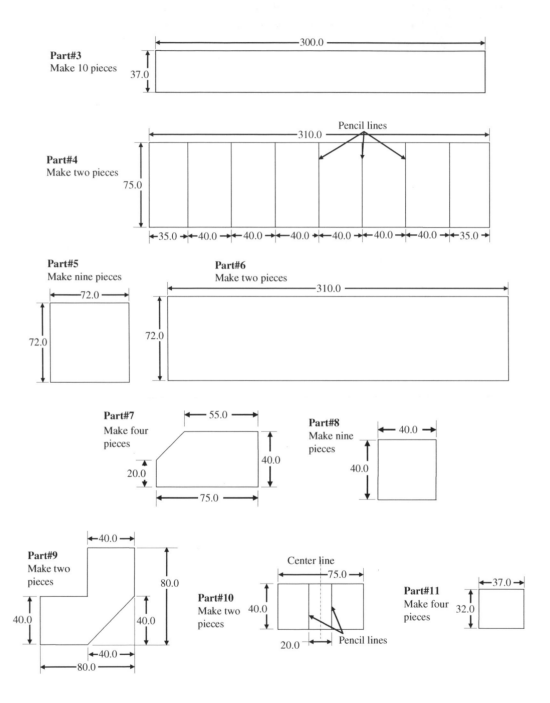

Part#3
Make 10 pieces

300.0

37.0

Part#4
Make two pieces

310.0

Pencil lines

75.0

35.0 40.0 40.0 40.0 40.0 40.0 40.0 35.0

Part#5
Make nine pieces

72.0

72.0

Part#6
Make two pieces

310.0

72.0

Part#7
Make four
pieces

55.0

40.0

20.0

75.0

Part#8
Make nine
pieces

40.0

40.0

Part#9
Make two
pieces

40.0

80.0

40.0

40.0

40.0

80.0

Part#10
Make two
pieces

Center line

75.0

40.0

20.0

Pencil lines

Part#11
Make four
pieces

37.0

32.0

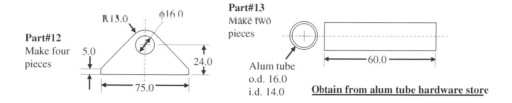

Part#12
Make four
pieces

R13.0   φ16.0

5.0

24.0

75.0

Part#13
Make two
pieces

Alum tube
o.d. 16.0
i.d. 14.0

60.0

Obtain from alum tube hardware store

12.2 **BASE STRUCTURE CONSTRUCTION**

Right let us start gluing up the base structure.

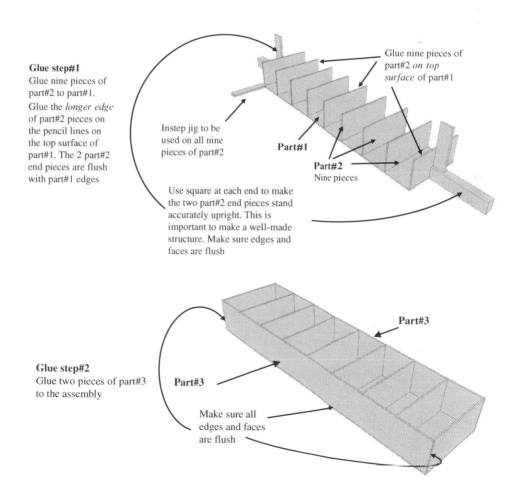

Glue step#1
Glue nine pieces of
part#2 to part#1.
Glue the *longer edge*
of part#2 pieces on
the pencil lines on
the top surface of
part#1. The 2 part#2
end pieces are flush
with part#1 edges

Instep jig to be
used on all nine
pieces of part#2

Glue nine pieces of
part#2 *on top
surface* of part#1

Part#1

Part#2
Nine pieces

Use square at each end to make
the two part#2 end pieces stand
accurately upright. This is
important to make a well-made
structure. Make sure edges and
faces are flush

Glue step#2
Glue two pieces of part#3
to the assembly

Part#3

Part#3

Make sure all
edges and faces
are flush

Glue step#3

Close the structural box with a second piece of part#1. Make sure all edges and faces are flush. Ensure you glue up the box on a flat surface. **You cannot untwist it after it is glued!**

Now make four more boxes

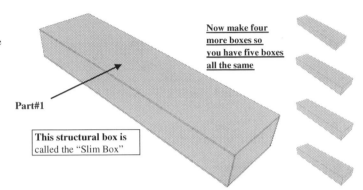

Now make four more boxes so you have five boxes all the same

Part#1

This structural box is called the "Slim Box"

Glue step#4

Now make a sixth box in the same style but this one is slightly longer and is fatter.

Glue nine pieces of part#5 to part#4. Once again glue on to the pencil lines and use the instep jig on all nine part#5 pieces. Use the square to keep the two ends upright. Its better to use two squares at one end to keep flush the edges and face.

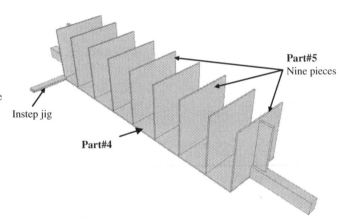

Part#5
Nine pieces

Instep jig

Part#4

Glue step#5

Glue two sides, part#6 on to the assembly. Make sure once again that all faces and edges are flush

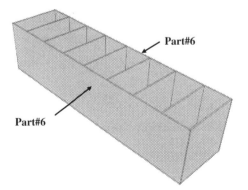

Part#6

Part#6

Glue step#6

Close the "fat" structural box by gluing the second part#4. Once again make sure that the gluing takes place on a flat surface so the box is not twisted.

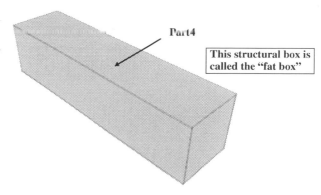

Part4

This structural box is called the "fat box"

Glue step#7

Now glue together one fat box and three slim boxes in the shape as shown here. Make sure you pay special attention to keeping the structural boxes *flat, square, parallel, and symmetrical.* You should discuss with classmates your strategy on how to do this. This is an important engineering layout problem in engineering, e.g., ship builders, aircraft makers, automobile constructors, bridge builders, house builders all have this problem. So now its your turn to work out how to do this. If you get it right then your job of assembling parts is less difficult. If you get it wrong then difficulty awaits you.

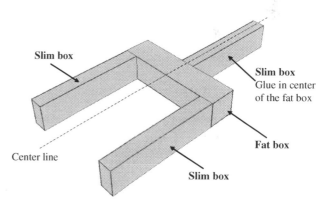

Slim box

Slim box
Glue in center of the fat box

Fat box

Slim box

Center line

Glue step#8

Now add two further slim boxes as shown. Once again you have a layout problem to solve that is three-dimensional rather than two-dimensional. This time you not only do you have to make your structure *flat, square, parallel, and symmetrical* but this time it has to be *upright and the two slim boxes have to be in the same plane when viewed from the side and parallel.* Good fun eh? By the way, sighting using the human eye is a very accurate way to check flatness.

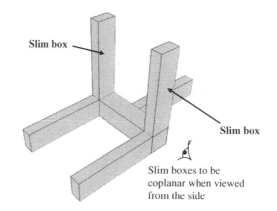

Slim box

Slim box

Slim boxes to be coplanar when viewed from the side

Glue step#9
The structure lacks strength at its glued faces so you need to glue reinforcement gussets. There are quite a few gussets so we start with four pieces of part#7. They need to be glued either side of the structure

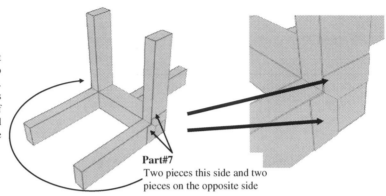

Part#7
Two pieces this side and two pieces on the opposite side

Glue step#10
Now glue six reinforcement pieces of part#8 on the underside of the structure. The three pieces on the ends of the slim boxes are reinforcement pieces for the rubber feet to be added soon.

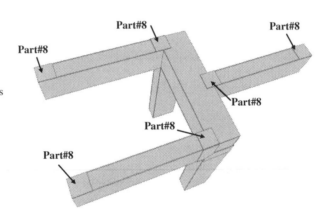

Part#8
Part#8
Part#8
Part#8
Part#8
Part#8

Glue step#11
Now add three more of part#8 reinforcement pads

Note that the reinforcement pads and gussets overlap equal amounts on the slim and fat boxes

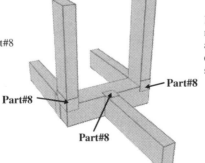

Part#8
Part#8
Part#8

Glue step#12
And finally the last of the reinforcement pieces; two pieces of part#9

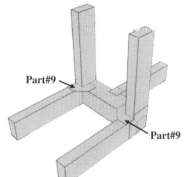

Part#9

Part#9 pieces are shaped to bridge across the fat and slim boxes

Part#9

Glue step#13
Make a small assembly that will be glued on to the base structure shortly. Glue two pieces of part#11 on to the pencil lines of part#10. Use the instep jig to set the pieces inboard. Also make sure the part#11 pieces are glued upright by using a square

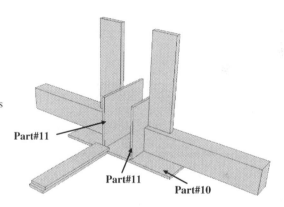

Part#11

Part#11

Part#10

Glue step#14
Now glue two pieces of part#12 on to the assembly

Part#12

Part#12

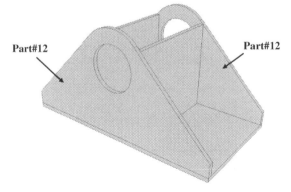

Glue step#14
Fit the aluminium tube into the assembly. It should be a little tight and sticking out either side by 10 mm.

You need another assembly so make one more

Part#13
Aluminium tube. The tube ends should be deburred.

Make one more assembly. You need two for the base structure

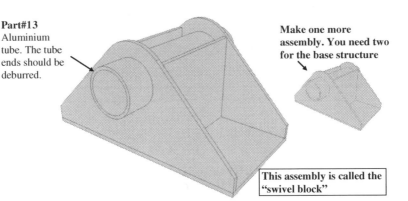

This assembly is called the "swivel block"

Glue step#15
Now glue the two swivel blocks on top of the vertical slim boxes of the structure. The assemblies are not "handed," meaning that there is no left hand or right hand assembly so just glue to match the top of the slim boxes. Make sure that the edges and faces are flush

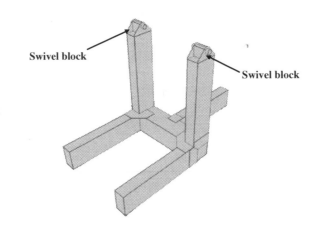

Swivel block

Swivel block

Final step for the base structure
Apply three pieces of stick-on rubber feet at each corner of the underside of the structure. The reinforcement squares glued on previously now provide for a more stable platform for the feet. You can get the feet from suppliers such as Parallax.com

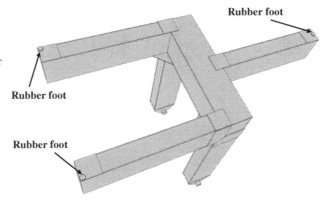

Rubber foot

Rubber foot

Rubber foot

Ok that's the base structure finished so let's move on to the drawings for the single-axis slideway which is in two parts. The first part is the slideway itself and the second part is the shuttle that slides on the slideway runners. Let's do the slideway first. Here are its drawings.

12.3 **SLIDEWAY DRAWINGS**

That is the end of the slideway drawings so let's start building it.

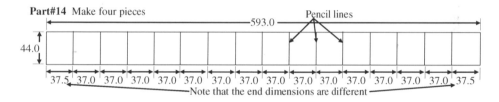

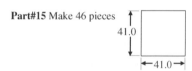

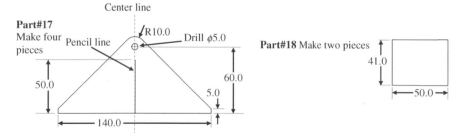

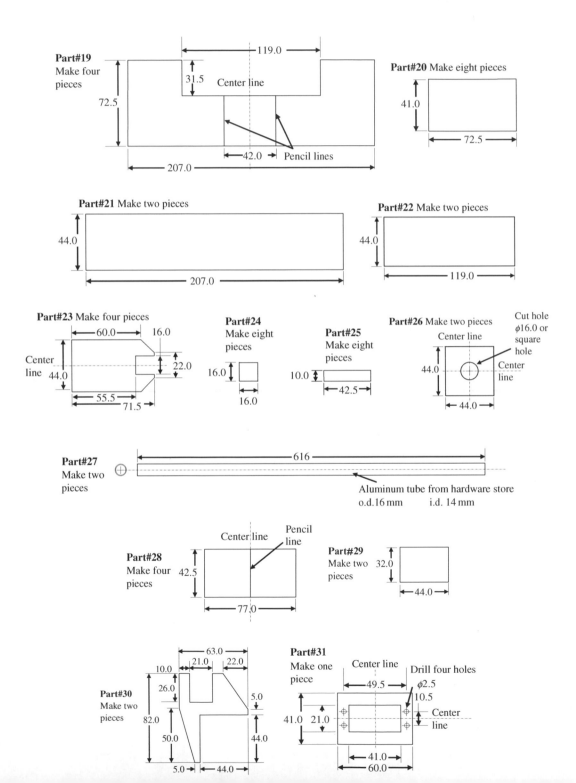

12.4 **SLIDEWAY CONSTRUCTION**

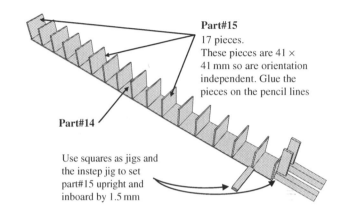

Glue step#1

Glue 17 pieces of part#15 to the top surface of part#14. Use the instep jig for all the part#15 pieces and make the two end pieces upright using the squares as jigs. The inside 12 pieces should be judged upright by eye

Part#15
17 pieces.
These pieces are 41 × 41 mm so are orientation independent. Glue the pieces on the pencil lines

Part#14

Use squares as jigs and the instep jig to set part#15 upright and inboard by 1.5 mm

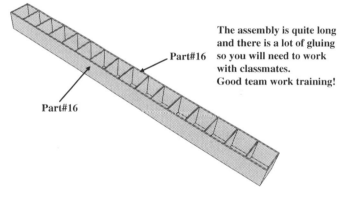

Glue step#2

Glue two pieces of part#16 on either side of the assembly. Make sure that edges and surfaces are flush

Part#16

Part#16

The assembly is quite long and there is a lot of gluing so you will need to work with classmates.
Good team work training!

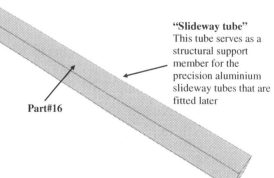

Glue step#3

Glue part#17, the top piece, to the slideway tube. Make sure that edges and surfaces are flush and make doubly sure that the box is not twisted. Remember it's the closing of the box, i.e., gluing the last piece, that determines if the box is twisted or not.

Part#16

"Slideway tube"
This tube serves as a structural support member for the precision aluminium slideway tubes that are fitted later

Glue step#4
Glue part#17 to the slideway tube.
This is a tricky operation. You
need to make the part (i) square
with the tube, (ii) flush with the
tube, and (iii) 195 mm from the end
of the tube. Work together to get
this right.

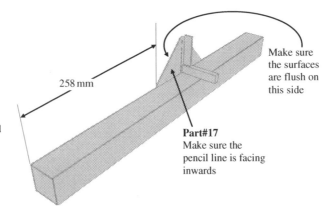

258 mm

Make sure
the surfaces
are flush on
this side

Part#17
Make sure the
pencil line is facing
inwards

90°

50.0.
The longer edge
is upright

Glue step#5
Glue part#18 to reinforce the
integrity of part#17

Part#18
Make sure the part
is at 90° across the
slideway tube

Glue step#6
Now glue another
part#17. Ensure that this
part is located accurately
either by checking the
edges are at 90° or by
measuring 258 mm as
before

**Temporarily put aside
these two "Slideway
tube" assemblies. Return
to them later**

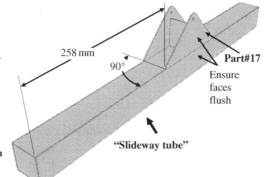

258 mm

90°

Part#17
Ensure
faces
flush

"Slideway tube"

**Now repeat glue steps
1–6 in order to make
another slideway tube
assembly**

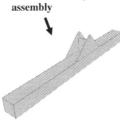

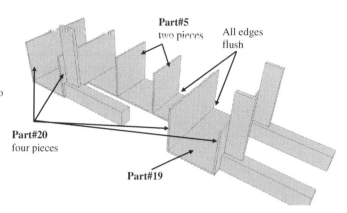

Glue step#7

Glue four pieces of part#20 and two pieces of part#15 on to the top surface of part#19. All the six pieces are upright by 41 mm. No need for the instep jig this time because all edges are flush

Part#5
two pieces

All edges flush

Part#20
four pieces

Part#19

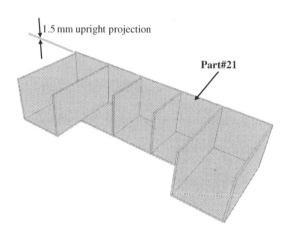

1.5 mm upright projection

Glue step#8

Glue part#21 to the assembly Make sure the edges and faces are flush on the side and bottom of the assembly which means that there will be an upright projection of 1.5 mm as shown

Part#21

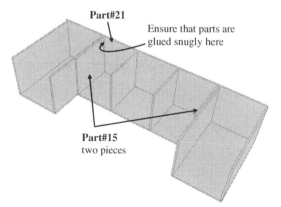

Part#21

Ensure that parts are glued snugly here

Glue step#9

Glue an additional two pieces of part#15 on to the inside faces of two pieces of part#20. Ensure that parts are glued snugly up against part#21

Part#15
two pieces

Glue step#10
Now glue another part#19 piece to the assembly. As usual make sure all edges and faces are flush

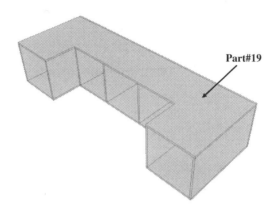

Part#19

Glue step#11
Now glue, part#22, the final part, into place.
....and you need to make another one. The assembly is called the **"slideway cross member"**

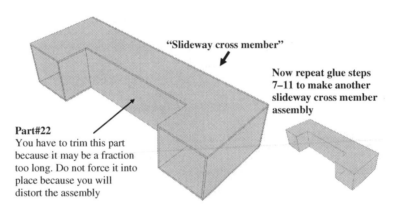

"Slideway cross member"

Now repeat glue steps 7–11 to make another slideway cross member assembly

Part#22
You have to trim this part because it may be a fraction too long. Do not force it into place because you will distort the assembly

Glue step#12
Now glue the two **slideway cross member** assemblies so as to connect the two **slideway tubes** to form the "slideway frame."
This gluing step should be done with great care and precision to make sure that the two slideway tubes are co-planar and that the arrangement forms a square and parallel rectangle when viewed from above. You need to discuss with your classmates how to do this. This construction, similar to the base structure built previously, is giving you valuable experience in the design and construction of mechanical structures.

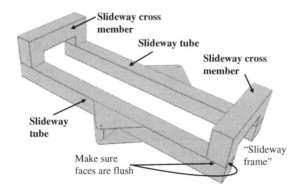

Slideway cross member

Slideway tube

Slideway cross member

Slideway tube

Make sure faces are flush

"Slideway frame"

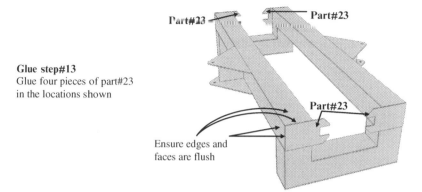

Part#23   Part#23

Glue step#13
Glue four pieces of part#23
in the locations shown

Part#23

Ensure edges and
faces are flush

Glue step#14
Now glue eight pieces of
part#24 on to the
assembly; two pieces at
each corner as shown.
These pieces are to
secure elastic bands
that will, in turn,
secure the precision
aluminum slideway
runners

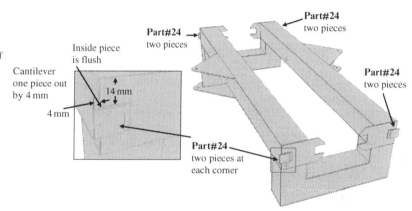

Part#24
two pieces

Part#24
two pieces

Part#24
two pieces

Inside piece
is flush

Cantilever
one piece out
by 4 mm

14 mm

4 mm

Part#24
two pieces at
each corner

Glue step#15
Glue eight pieces of
part#25 on the side of
the slideway frame.
Be extremely careful
to position these parts
accurately to the
dimensions shown
**and in the longer
section of the
slideway frame**

Note the overlapping technique
of building the square. Equal
length sided square with four
equal length pieces

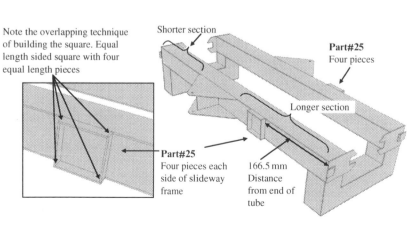

Shorter section

Part#25
Four pieces

Longer section

Part#25
Four pieces each
side of slideway
frame

166.5 mm
Distance
from end of
tube

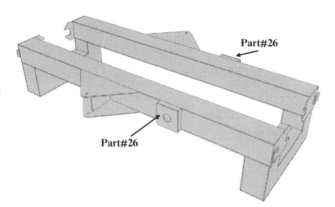

Glue step#16
Glue the square piece part#26 on to the four pieces of part#25 making sure that faces and edges are flush. Now repeat on the other side

Part#26

Part#26

Slot the two pieces of aluminum tube, part#26 in to the slots of part#23 and secure with elastic bands as shown. The aluminum tubes should over hang 10 mm each end of the slots

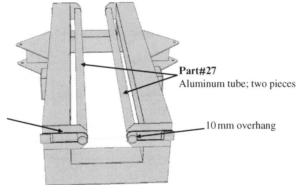

Part#27
Aluminum tube; two pieces

Elastic band

10 mm overhang

Glue step#17
Glue three pieces of part#15 on to top surface of part#28. The middle piece is to align with the pencil mark. The three pieces are to be inboard by 1.5 mm using the instep jig. The parts will be flush the other side. Use squares as jigs to make the two end pieces upright

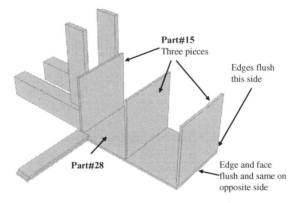

Part#15
Three pieces

Edges flush this side

Part#28

Edge and face flush and same on opposite side

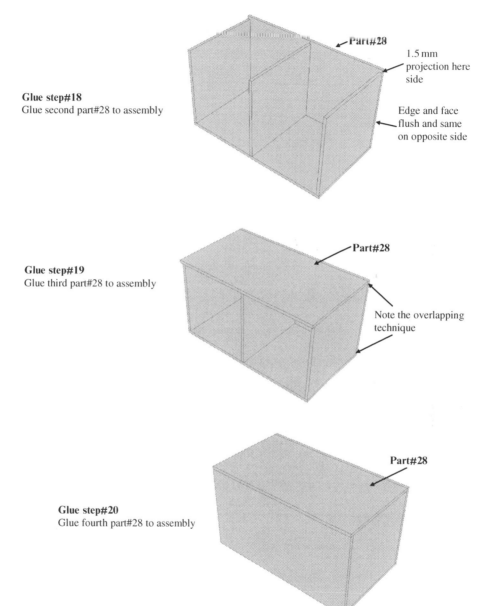

Glue step#18
Glue second part#28 to assembly

Part#28

1.5 mm projection here side

Edge and face flush and same on opposite side

Glue step#19
Glue third part#28 to assembly

Part#28

Note the overlapping technique

Glue step#20
Glue fourth part#28 to assembly

Part#28

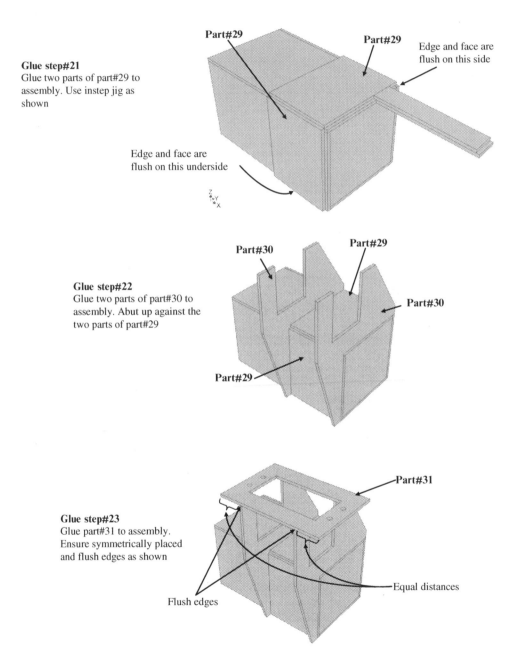

Glue step#21
Glue two parts of part#29 to assembly. Use instep jig as shown

Part#29

Part#29

Edge and face are flush on this side

Edge and face are flush on this underside

Glue step#22
Glue two parts of part#30 to assembly. Abut up against the two parts of part#29

Part#30

Part#29

Part#30

Part#29

Glue step#23
Glue part#31 to assembly. Ensure symmetrically placed and flush edges as shown

Part#31

Flush edges

Equal distances

Glue step#24

"Pull back servo"

Secure Futaba BLS177SV servo with 4 M3 screws stainless steel pan crosshead 8 mm long. Carefully self tap the screws and do not tighten too much. Also use M3 washers. This is a high-performance high torque servo (3.7 N m). The servo is called the **"pull-back servo."** A low-performance servo cannot be used because a force of 20 N is required to pull-back the springs to maximum displacement. The torque arm of the pull-back system is approximately 0.1 m in length so 2 N m is required to pull back the ball shuttle. The additional remaining torque of the servo in reserve can be used to uprate the catapult system

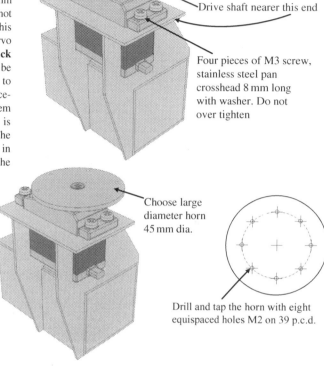

Drive shaft nearer this end

Four pieces of M3 screw, stainless steel pan crosshead 8 mm long with washer. Do not over tighten

Screw on a large 45 mm diameter horn. You will need eight screw holes drilled

Choose large diameter horn 45 mm dia.

Drill and tap the horn with eight equispaced holes M2 on 39 p.c.d.

Glue step#25

Glue the assembly on to the slideway frame. Make sure you glue it on the correct end and the correct corner of the frame as shown. You may find it better to remove the heavy servo before gluing

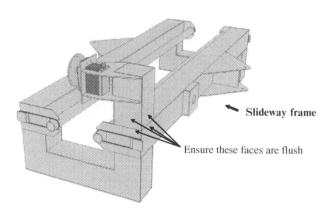

Slideway frame

Ensure these faces are flush

That's the end of the construction of the slideway frame. Now let's move on to the construction of the shuttle and pull-back system.

12.5 **SHUTTLE AND PULL-BACK SYSTEM DRAWINGS**

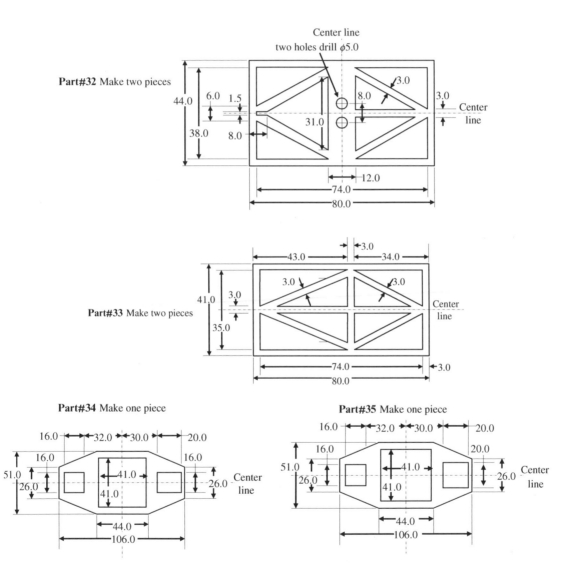

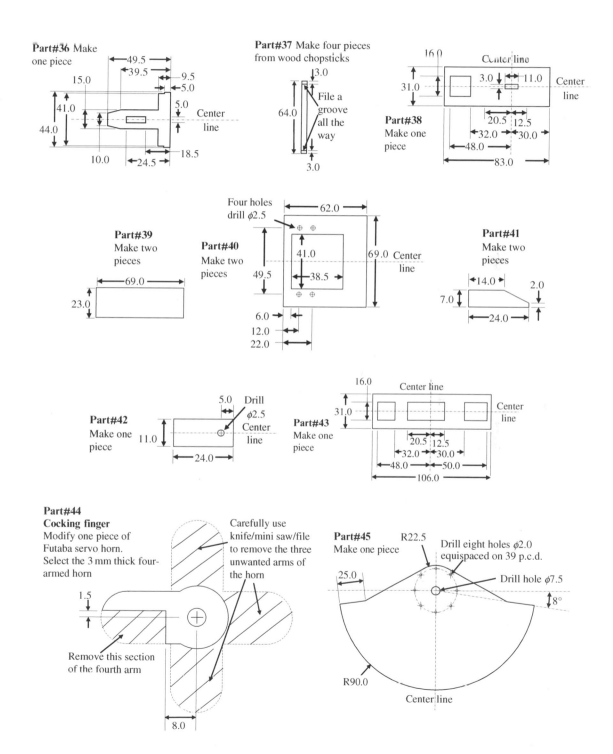

Part#36 Make one piece
49.5
39.5
15.0
9.5
5.0
5.0
41.0
Center line
44.0
18.5
10.0
24.5

Part#37 Make four pieces from wood chopsticks
3.0
File a groove all the way
64.0
3.0

16 0
Center line
3.0
11.0
Center line
31.0
Part#38 Make one piece
20.5 12.5
32.0
30.0
48.0
83.0

Part#39 Make two pieces
69.0
23.0

Part#40 Make two pieces
Four holes drill ⌀2.5
62.0
41.0
69.0 Center line
49.5
38.5
6.0
12.0
22.0

Part#41 Make two pieces
14.0
2.0
7.0
24.0

Part#42 Make one piece
5.0 Drill ⌀2.5
Center line
11.0
24.0

Part#43 Make one piece
16.0
Center line
31.0
Center line
20.5 12.5
32.0 30.0
48.0 50.0
106.0

Part#44
Cocking finger
Modify one piece of Futaba servo horn.
Select the 3 mm thick four-armed horn
Carefully use knife/mini saw/file to remove the three unwanted arms of the horn
1.5
Remove this section of the fourth arm
8.0

Part#45 Make one piece
R22.5
25.0
R90.0
Drill eight holes ⌀2.0 equispaced on 39 p.c.d.
Drill hole ⌀7.5
8°
Center line

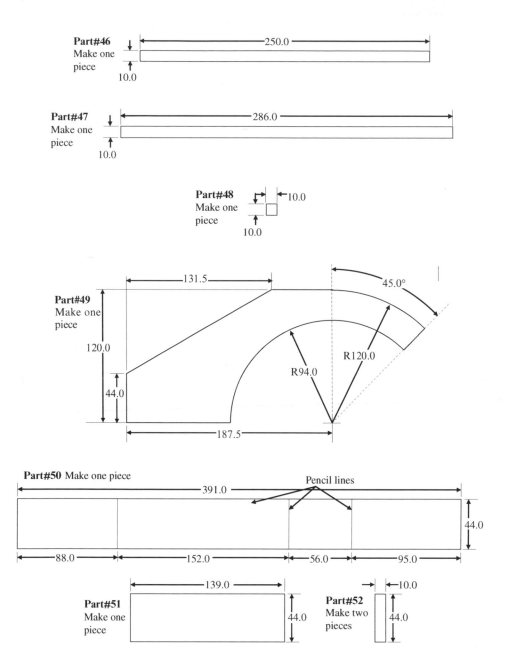

Ok that's all the drawing parts done for the shuttle and pull-back system.
Now, let's move on their construction.

12.6 **CONSTRUCTION OF SHUTTLE AND PULL-BACK SYSTEM**

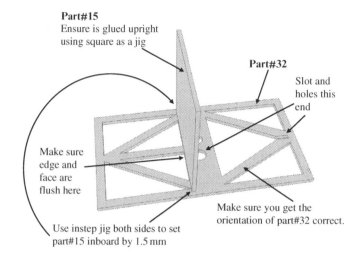

Glue step#1
Glue part#15 to part#32. Use a square to make part#15 upright and use the instep jig to set inboard part#15 by 1.5 mm both sides

Part#15
Ensure is glued upright using square as a jig

Part#32
Slot and holes this end

Make sure edge and face are flush here

Make sure you get the orientation of part#32 correct.

Use instep jig both sides to set part#15 inboard by 1.5 mm

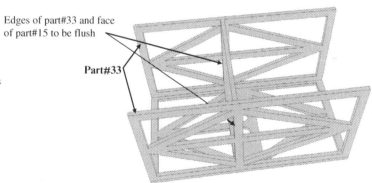

Glue step#2
Glue two pieces of part#33. Make sure flush edges and faces as shown

Edges of part#33 and face of part#15 to be flush

Part#33

Glue step#3
Glue another piece of part#32 on top of the assembly. Make sure you get its orientation correct. Also be careful to get all edges and faces flush. This assembly is very lightweight and flimsy so take care.

Part#32

The two pairs of holes are mirror-imaged

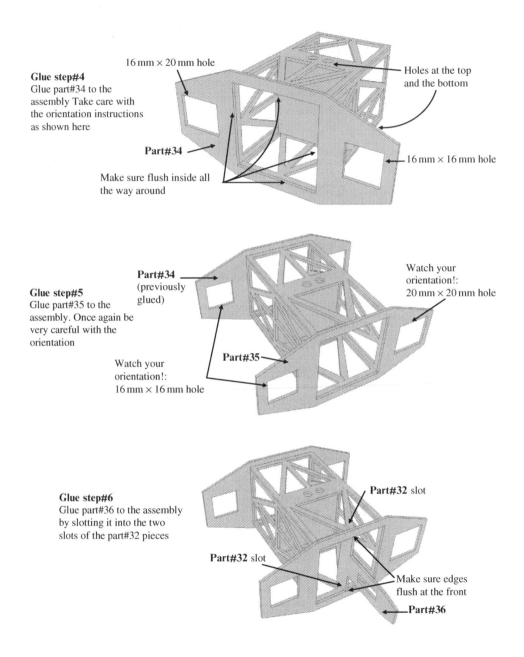

16 mm × 20 mm hole

Holes at the top
and the bottom

Glue step#4
Glue part#34 to the
assembly Take care with
the orientation instructions
as shown here

Part#34

16 mm × 16 mm hole

Make sure flush inside all
the way around

Part#34
(previously
glued)

Watch your
orientation!:
20 mm × 20 mm hole

Glue step#5
Glue part#35 to the
assembly. Once again be
very careful with the
orientation

Part#35

Watch your
orientation!:
16 mm × 16 mm hole

Part#32 slot

Glue step#6
Glue part#36 to the assembly
by slotting it into the two
slots of the part#32 pieces

Part#32 slot

Make sure edges
flush at the front

Part#36

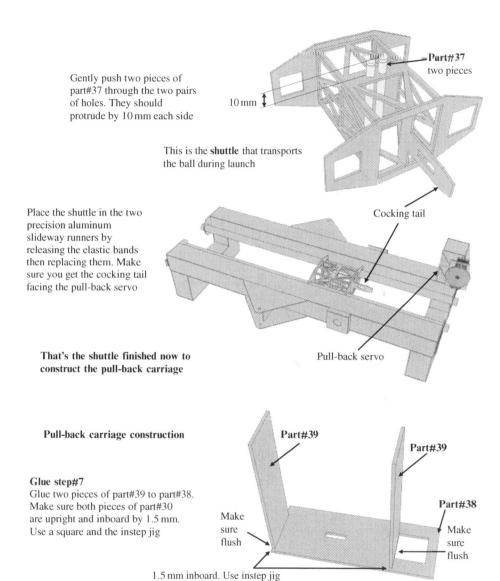

Gently push two pieces of part#37 through the two pairs of holes. They should protrude by 10 mm each side

10 mm

Part#37
two pieces

This is the **shuttle** that transports the ball during launch

Cocking tail

Place the shuttle in the two precision aluminum slideway runners by releasing the elastic bands then replacing them. Make sure you get the cocking tail facing the pull-back servo

Pull-back servo

That's the shuttle finished now to construct the pull-back carriage

Pull-back carriage construction

Part#39

Part#39

Glue step#7
Glue two pieces of part#39 to part#38. Make sure both pieces of part#30 are upright and inboard by 1.5 mm. Use a square and the instep jig

Make sure flush

Part#38

Make sure flush

1.5 mm inboard. Use instep jig

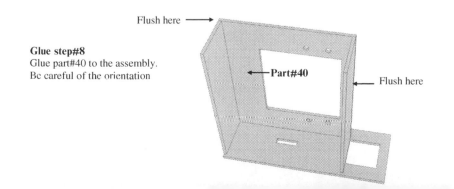

Flush here

Glue step#8
Glue part#40 to the assembly. Be careful of the orientation

Part#40

Flush here

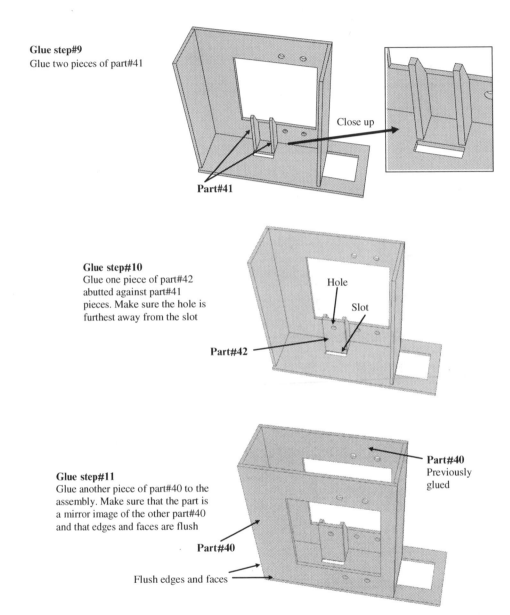

Glue step#9
Glue two pieces of part#41

Close up

Part#41

Glue step#10
Glue one piece of part#42
abutted against part#41
pieces. Make sure the hole is
furthest away from the slot

Hole

Slot

Part#42

Glue step#11
Glue another piece of part#40 to the
assembly. Make sure that the part is
a mirror image of the other part#40
and that edges and faces are flush

Part#40
Previously
glued

Part#40

Flush edges and faces

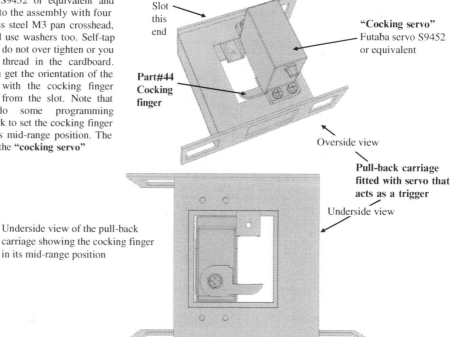

Glue step#12
Glue part#43 to the assembly. Be very careful to get the orientation correct. You get this by matching the 16 mm × 16 mm holes on one side

16 mm × 20 mm hole

Part#43

Edge and face flush here

16 mm × 16 mm holes matched on this side

Now fit part#44 cocking finger to a Futaba servo S9452 or equivalent and then secure it to the assembly with four screws stainless steel M3 pan crosshead, 8 mm long and use washers too. Self-tap the screws and do not over tighten or you will strip the thread in the cardboard. Make sure you get the orientation of the servo correct with the cocking finger furthest away from the slot. Note that you must do some programming calibration work to set the cocking finger as shown in its mid-range position. The servo is called the **"cocking servo"**

Slot this end

"Cocking servo"
Futaba servo S9452 or equivalent

Part#44 Cocking finger

Overside view

Pull-back carriage fitted with servo that acts as a trigger

Underside view

Underside view of the pull-back carriage showing the cocking finger in its mid-range position

Now temporarily remove the elastic bands and enter the pull-back carriage into the aluminum slideway runners. Make sure the orientation is correct with the slot at the rear. Both the shuttle and the pull-back carriage should slide easily and freely. If not you have to inspect and rectify the problem. Engineering design involves a lot of fault finding. You need to be diligent and patient.

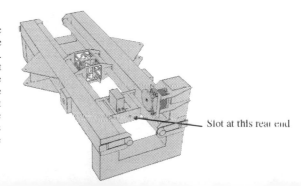

Slot at this rear end

Now fit eight springs as four pairs of springs mounted symmetrically above and below the shuttle. Each pair of springs is two Misumi AUA-6-60 stainless steel springs hooked together in series.

Two Misumi AUA-6-60 stainless steel springs hooked together in series.

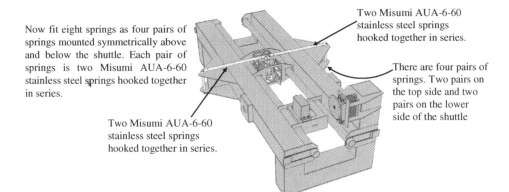

Two Misumi AUA-6-60 stainless steel springs hooked together in series.

There are four pairs of springs. Two pairs on the top side and two pairs on the lower side of the shuttle

You should check that the shuttle cocking tail slides easily into the pull-back carriage and then you can play by manually moving the cocking finger to engage in the slot of the cocking tail. This gives you the idea of how the pull-back system operates manually then later you will program the operation

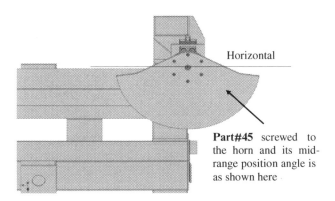

Cocking tail slot

Cocking finger

Cocking tail

Now screw part#45 onto a large 45 mm diameter horn and secure with 6 pieces of M2 screw, stainless steel pan cross head, 6 mm long. If there is no large diameter horn available then use a smaller diameter horn. Its just that the bigger the horn diameter, the better. Now set the mid-range servo position as shown here in the diagram

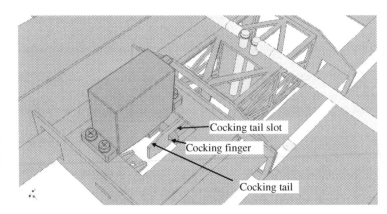

Horizontal

Part#45 screwed to the horn and its mid-range position angle is as shown here

Glue step#13
Glue curved strip part#46 on to the rim of part#45 so that the curved strip *is centrally placed* on to the rim of part#45. There is a 10 mm gap at the top of the curved rim. This operation needs patience to do a quality job. Part#45 and part#46 form the **"pull-back arc"**

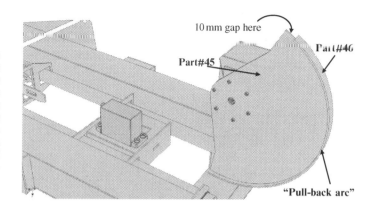

10 mm gap here

Part#46

Part#45

"Pull-back arc"

Glue step#14
Glue small piece part#48 at the furthest end of the long thin part#47. The assembly forms the **"push-pull strip"**

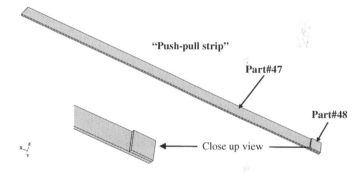

"Push-pull strip"

Part#47

Part#48

Close up view

Glue step#15
Now glue the push-pull strip to the pull-back arc but only glue the first 10 mm! Do not use too much glue and make sure the curved strip follows the rim of the pull-back arc

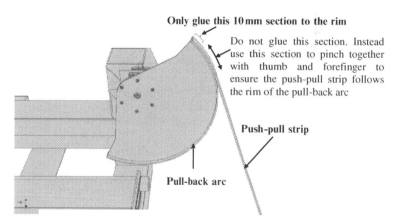

Only glue this 10 mm section to the rim

Do not glue this section. Instead use this section to pinch together with thumb and forefinger to ensure the push-pull strip follows the rim of the pull-back arc

Push-pull strip

Pull-back arc

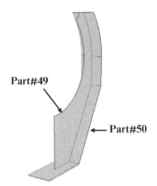

Glue step#16
Glue part#50 to part#49 such that part#49 sits mid-range of part#50 as shown in the diagram

Part#49

← **Part#50**

This assembly will keep the push-pull strip in close conformity with the pull-back arc. Without the assembly, the push-pull strip would not be able to push. It would only be able to pull

Glue step#17
Glue the assembly on the back of the cross member assembly. Make sure you glue it on centrally

Glue step#18
Glue one piece of part#51 and two pieces of part#51 underneath the slideway tubes. This is a bridge that inhibits the push-pull strip from bending too much when pushing the pull-back carriage. Decide on a suitable location about half way between the carriage and the cross member.

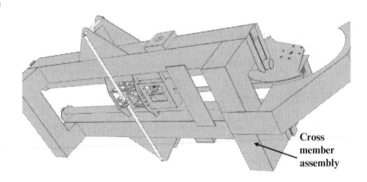

Cross member assembly

Slot in pull-back carriage

Now feed the push-pull strip into the slot of the pull-back carriage....

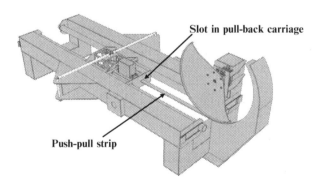

Push-pull strip

....and secure the push-pull strip inside the pull-back carriage with an M3 stainless steel screw pan cross head 5 mm long. You access the screw hole on the underside of the pull-back carriage

Underside view of pull-back carriage

Screw securing push-pull strip

Rotate the pull-back arc

You should play with writing code now to rotate the pull-back arc and to engage the cocking finger so as to pull back the ball carriage then to let go the carriage

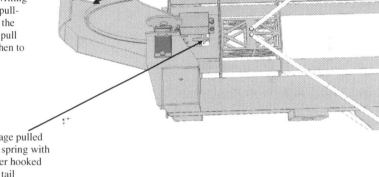

Here is the carriage pulled back against the spring with the cocking finger hooked into the cocking tail

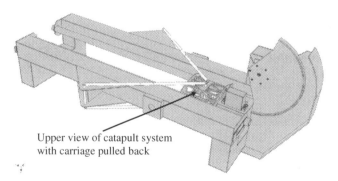

Upper view of catapult system with carriage pulled back

12.7 **ASSEMBLY OF COMPLETE ROBOT**

Now locate the catapult system into the base structure. The design of the angle drive system is left to the student. We will give here photographs of the solution. A Futaba, S5801 "winch" servo is used together with a capstan drum. The cable length is nonlinear as the servo drives through the angle range so springs accommodate the change in cable length and also provide tension in the cable so as to make the capstan effective. Without cable tension capstans are useless.

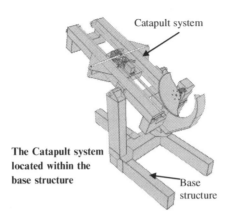

Catapult system

The Catapult system located within the base structure

Base structure

Side view of the catapult robot showing the cable system that gives elevation angle to the catapult launching velocity

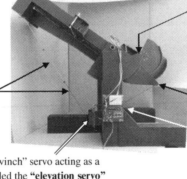

Cable with two Misumi AUA-6-60 springs hooked in series. The springs keep the cable in tension. The cable is wrapped around a capstan drum

Futaba S5801 "winch" servo acting as a capstan. It is called the **"elevation servo"**

This is an **elevation angle sensor** that you can read by eye. It is made from cardboard and has pencil markings to give the catapult elevation angle to ½ ° angle resolution

Other end of cable is tied to this cantilever

The Basic Stamp USB Board of Education is screwed to the **slim box** of the base structure. The Basic Stamp microcomputer controls just three servos which are:
(i)   The pull-back servo
(ii)   The cocking servo
(iii)   The elevation servo

Here is a view of the Futaba winch servo driving a 30 mm diameter drum made from cardboard

The drum is made from cardboard and its diameter is 30 mm. The manufacture of the drum from cardboard and its attachment to the horn is an interesting exercise for students

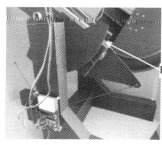

These two photographs, left and right, show more details of the elevation angle sensor and the capstan servo

Elevation angle sensor

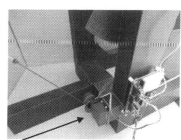

Capstan winch servo

That completes the construction of the Catapult robot. Before we turn our attention to its programming we give a brief introduction the "winch" servo.

12.8 THE "WINCH" SERVO

Fig. 12.1 shows an inside view of a winch servo. It differs on two counts with the standard servo described in Chapter 7.

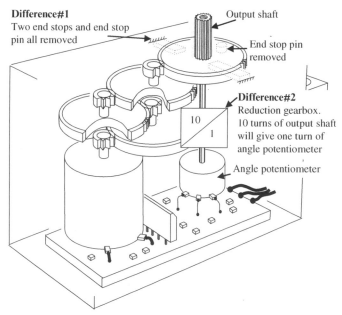

Difference#1
Two end stops and end stop pin all removed

Output shaft

End stop pin removed

Difference#2
Reduction gearbox. 10 turns of output shaft will give one turn of angle potentiometer

10 / 1

Angle potentiometer

■ **FIG. 12.1** Inside a winch servo showing the two differences between the standard servo described in Chapter 7.

Difference#1 is that the two end stops and the end stop pin are removed. This means that the output shaft can continuously rotate.

Difference#2 is that there is a 10:1 reduction gearbox, usually a worm and wheel, between the output shaft and the angle potentiometer that measures angle. This means that if the output shaft rotates by 10 turns, then the angle potentiometer will measure just one turn.

Quite simply the gearbox and potentiometer fool the output shaft into thinking that the output shaft has only rotated by 1/10th of what it thinks so the output shaft will rotate 10 times more than a 1:1 gearbox.

The winch servo control is the same as the standard servo. A train of 1 ms pulse widths will set the servo to 0 degrees position and a train of 2 ms pulses will drive the servo to 3600 degrees. The winch servo is useful if you want more than the standard 140 degrees rotation and also the winch servo can do more than 20 times the work of the standard servo because work done is equal to the product of torque and angle rotated.

There is a further important feature of the winch servo and that concerns a small gain potentiometer secreted in the side of the servo housing that can be rotated with a special screwdriver that comes supplied in its delivery box. This gain potentiometer amplifies the angle potentiometer output to give an electrical, rather than mechanical, fooling of the position of the output shaft. It means that by varying the gain potentiometer the output shaft can rotate by 3 turns or by 10 turns for a 1–2 ms signal depending on the tweaking of the gain potentiometer. This is useful if, say, for example, 10 turns is too much rotation. Yes, the number of turns could be set by a software pulse width limit but you obtain higher angle resolution by using the whole pulse width range from 1 to 2 ms.

There is a down side to the winch servo. It is the backlash in the worm and wheel reduction gearbox that is a form of hysteresis. It means that the servo will have a dead zone of a few degrees at a set position. You can decrease the effect of this problem by always positioning the servo in one direction, say going CW. So if you want to position the servo more CCW than its current position, then drive the servo a few degrees too much CCW from the required position then approach the required position in a CW direction.

The author notes that a Chinese manufacturer, JX Servo is producing winch servos and standard servos that are a lot cheaper than Futaba servos. The performance is also indistinguishable. These servos are obtainable from, for example, Chinese websites, Taobao, Aliexpress, or from Hobbyking in the United States.

Alright, that concludes our discussion of the electromechanical construction of the Catapult robot. We now turn out attention to its real-time programming.

12.9 PROGRAMMING THE CATAPULT TO LAUNCH A BALL AUTOMATICALLY

```
'Programme for Basic Stamp2 microcomputer
'Programme for the catapulting one ping pong ball that has been
manually loaded into the ball carriage.

pullbackservo    VAR    WORD    'variable that represents displacement of ball carriage against spring
                                force
cockingservo     VAR    WORD    'variable that represents engage or release of ball carriage
launchanglepin   con    14      'the pin connection to the winch servo that elevates the catapult system
pullbackpin      con    13      'the pin connected to the pull back servo
cockingpin       con    12      'the pin connected to the cocking servo that engages or releases ball
                                carriage
'The following numbers are calibrated values for the servos found by experiment
'425=forward ready cock, this is the angle for the pull back servo to go forward to engage with carriage
'1098=full pullback, this is the angle for the pull back servo for maximum launch speed
'500=release,780=cock, these are angles for the cocking servo to disengage and engage with carriage
i        var       word 'counting variable
pullbackservo=1098 'arrange for pullback carriage to be in starting position
                 PAUSE 500              'wait for ½ second
'start of main programme now
'drive pull back carriage forward to engage with ball carriage and engage with cocking finger
                 FOR i=1 to 25    'give about 25 x 20ms=0.5 sec for this operation
                 PULSOUT pullbackpin, 425    'pull back carriage go forward
                 PULSOUT cockingpin, 780     'engage cocking finger with ball carraige
                 PAUSE 15                     'delay is requirement for the servos
                 NEXT
'keep the cocking finger engaged and simultaneously pull back the ball carraige
                 FOR i=1 to 50      ' give about 50 x 20ms = 1sec for this operation
                 PULSOUT pullbackpin,pullbackservo   'drive to full pull back displacement
                 PULSOUT cockingpin,780 'keep the cocking finger engaged
                 PAUSE 15                      'delay is requirement for servos
                 NEXT
'pulled back and catapult the ball
                 FOR i=1 to 50    'give about 50 x 20ms = 1sec for this operation
                 PULSOUT pullbackpin,pullbackservo   'keep the carriage pulled back
                 PULSOUT cockingpin,500   'release the cocking finger thus catapulting the ball
                 PAUSE 15                       'delay is a requirement for the servos
                 NEXT
```

```
'drive pullback carriage forward with cocking finger disengaged ready for new operation
                 FOR i=1 to 50    'give about 50 x 20ms = 1sec for this operation
                 PULSOUT pullbackpin,425    'forward position for the pullback carriage
                 PULSOUT cockingpin,500     'keep the cocking finger in release position
                 PAUSE 15                   'delay is requirement for the servos
                 NEXT
again1:          GOTO again1                'infinite loop to stop the programme
```

We conclude this book with some interesting problems some of which are highly advanced and challenging.

12.10 **PROBLEMS**

1. Design and build a sensor system that checks to see if the cocking system is engaged properly. In other words to make sure that the cocking finger arm is fully engaged in the slot.

2. The cocking finger will damage the cocking slot after some time of operation so you will need to check the location of the slot damage and use a hard material such a short length of tooth pick. This will also decrease ball-in-basket repeatability errors. A very interesting exercise is to use a camera to video the cocking finger as it lets go of the cocking tail to see what happens to the cocking tail in slow motion.

3. For higher launching speed accuracy, it may be necessary to measure the pull-back displacement to a resolution of maybe 0.5 mm, resolution for you to decide. How this could be done?

4. Design a catapult elevation angle system using a stepper motor or a winch servo that rotates the capstan in lieu of the Futaba winch servo. Remember a stepper motor has an unknown position on power-up so you will need to datum the catapult system position with a sensor, see Section 6.5 in Chapter 6.

5. Derive a polynomial equation that gives the winch servo angle as a function of desired elevation angle of the catapult system. It is only necessary for the equation to give 0.5 degrees elevation angle accuracy in the first instance. It may be necessary to achieve greater accuracy for shots that are fired near horizontal or near vertical. The winch servo has backlash, otherwise known as hysteresis, is that may mean that 0.5 degrees accuracy is only possible if you drive the servo increasing to the desired angle or decreasing to the desired angle.

6. Design and construct a waste paper basket goal mounted such that the mouth of the basket is set at 1 m height. Improve the design with a hole in the bottom of the basket that allows the ball to fall through and to be counted. Link this to a large scoreboard that can be seen across the room for everybody to see.

7. Design a mounting system that fixes the 10-ball magazine autoloader, featured in Chapter 8, to the catapult system so that the catapult robot is capable of rapid fire.

8. Alternatively and more advanced, don't use Chapter 8 autoloader. Instead, much more interestingly, design a mounting system to place a metal mesh waste paper bin on top of the robot such that the basket mouth is 1 m above the table surface at the same height of a basket goal set up for robot ping-pong basketball. Now build a ball delivery system built in to the bottom of the basket such that 10 balls can be stored and delivered one at a time, on demand, to the shuttle. Such a robot forms the basis of a fascinating game of ping-pong basketball where you can pass ball from robot to robot then shoot for a goal. There is no need to complicate matters with a ball picker-upper because being able to store 10 balls means that if you miss a pass, then you have stored balls to keep the game going. Devising a set of rules, a scoring system and penalty loss methodology is also very interesting.

9. Largely redesign the catapult system to eliminate recoil. This is a major project and it has the advantage of no shaking shock being given to the base structure. A shock-less system is likely to give a more accurate catapult. One solution is to have masses connected to strings that run over pulleys such that (i) the masses travel in equal displacements to the shuttle but in opposite directions and (ii) the opposite moving masses equal the mass of the shuttle plus, say, half the mass of the ball. The catapult will be largely recoilless but the down side is that half the spring energy is dissipated by the opposite moving masses. So to achieve 12 m/s launch speed, you will need to increase the rate of the springs and increase the torque rating of the pull-back servo. Carry out a YouTube check to see what other people are doing along these lines, for example, check out latest developments in archery

10. Can a sixth order (or more or less order) polynomial equation be fitted to the x-y coordinates of the ball trajectory such that the coefficients of each term of the polynomial are a function of the L_H and L_R values, Chapter 2, Fig. 2.14. This equation can be used, that given L_H and L_R, the elevation angle and pull-back spring displacement can be easily computed.

11. Design a human eye target aimer that computes L_H and L_R values, Chapter 2, Fig. 2.14. Then improve the system to develop an autonomous artificial intelligent agent system that uses a smart phone attached to the catapult system that sets the catapult system elevation angle and pull-back spring displacement to catapult the ball into the basket. The human still has to direct the catapult toward the basket, so....next problem to solve is....

12. Put omnidirectional wheels on to the base structure such that the catapult robot can run around on a game table and can direct the catapult system in the direction of the basket by steering about a vertical axis.

Index

Note: Page numbers followed by "*f*" indicate figures.

Printed in the United States
By Bookmasters